AI is not just transforming mental health—it is redefining the role of psychologists in the digital age. As technology reshapes how we provide assessment, diagnosis, and engagement with clients (and each other), we must take an active role in guiding its integration (via advocacy, policymaking, and oversight). Psychologists should be part of the team of professionals who contribute to the future pathways of AI. Beyond ethical and regulatory concerns, AI challenges us to reimagine therapeutic relationships (with clients, colleagues, and systems), clinical decision-making, and the future of professional practice. This book provides the insights and strategies psychologists need to harness AI responsibly ensuring we lead with expertise, uphold the integrity of our field, and maximize AI's potential to enhance delivery of cutting-edge quality of care. Chapter 4 outlines an incredible wealth of knowledge of the utilization of AI in clinical practice and Chapter 10 provides the necessary ethical considerations for the appropriate integration of AI into practice. This book will have an ineffable impact in our field, and I highly recommend it.

— Joshua Heitzmann, PhD, 2025 President of the California Psychological Association

This book strikes the right balance: it welcomes AI into the therapy room as a helpful tool, while firmly keeping the therapist—and the therapeutic relationship—at the center of the work.

— Luis Maimoni, MS, LMFT

As AI integrates into mental health care, psychologists must stay informed to protect the integrity of our profession. AI models, if left unchecked, can reinforce systemic biases and misinterpret cultural nuances, leading to misdiagnoses and treatment inequities. Understanding AI's ethical and regulatory challenges is no longer optional—it's essential. This book equips psychologists with the knowledge to engage AI responsibly and advocate for its ethical use in therapy and assessment.

— Dr. Rivka Edery, Psy.D., L.C.S.W.

AI is no longer an emerging concept; it is becoming part of the infrastructure of healthcare. In fields like psychology, where empathy, ethics, and context are essential, we need professionals who understand how AI works, where it fits, and when it does not. This book is an important step toward building that understanding.

— Timothy Chou, PhD, and CEO of BevelCloud.ai

Creativity is about forming connections between experiences, memories, and ideas, transforming them into novel insights through imagination. As an expressive arts therapist, I see AI as a tool that enhances this process, opening new pathways for exploration and self-expression. To harness AI's full potential in therapy, mental health professionals would do well to understand its role and ethical integration. This book provides essential guidance, ensuring AI becomes a powerful ally in therapeutic practice.

— Lindsey Sherwin, M.Sc., RDT, EXAT

AI is rapidly reshaping psychology, and those who fail to engage with it risk falling behind. As mental health professionals, we must evolve alongside technology to remain relevant and effective. Clients will increasingly expect AI-assisted tools in therapy, and research institutions are already integrating AI-driven methodologies. This book is a must-read for clinicians who want to stay competitive and lead the future of mental health, rather than struggle to keep up.

— Anna-Christina Jackson, MBACP MNCS (Accred)

Artificial Intelligence
for
Mental Health
Professionals

Other Books by A. Vincent Vasquez, MS, MBA

- *Relationship Workout: The Men's Manual*
  A practical, no-nonsense guide for men to build stronger, healthier intimate relationships—structured around a 5-step Relationship Workout Plan and focused on helping men develop stronger relationship skills in 12 core areas. Foreword by Brenda Hart, PhD.

- *Relationship Workout: Master Class Coursebook*
  An in-depth companion to the Men's Manual, this is the coursebook for the Relationship Workout Master Class, designed to help men actively strengthen their relationship skills on a journey to having more fun and less drama in their intimate relationships. Foreword by Brenda Hart, PhD.

- *The Next CIO*
  A forward-looking guide for technology leaders, this book explores how Enterprise Technology Management (ETM) enables CIOs to lead more autonomous, efficient, and strategically aligned IT organizations. Foreword by David Cappuccio.

- *Precision Construction*
  A practical framework for applying Internet of Things (IoT) technologies to modern construction, helping industry professionals improve efficiency, accuracy, and project outcomes. Foreword by Matthew Flannery, MS.

Artificial Intelligence for Mental Health Professionals

Brenda Hart, Ph.D.
Antonio Diego Vasquez
A. Vincent Vasquez, MS, MBA

CROWDSTORY

CROWDSTORY

Relationship Workout

Published in partnership with Relationship Workout, LLC
Visit us at: RelationshipWorkout.com

For additional resources related to this book, including continuing education (CE) materials and credits, visit: RelationshipWorkout.com/CE

ISBN: 979-8-9993086-0-3

Edition 2.3 Build 33

Contents

FOREWORD X

PREFACE XV

PEER REVIEW ACKNOWLEDGMENT XVI

CHAPTER 1: INTRODUCTION 1

WHY THIS BOOK, AND WHY NOW? 1

WHO THIS BOOK IS FOR 2

HOW THIS BOOK IS STRUCTURED 3

HOW TO USE THIS BOOK 4

DISCLAIMER ON COMPANY AND PRODUCT MENTIONS 4

CONTINUING EDUCATION CREDITS AND CONTENT UPDATES 5

A FINAL WORD BEFORE WE BEGIN 6

CHAPTER 2: DEMYSTIFYING AI 7

HOW AI LEARNS 8

HOW AI GENERATES ANSWERS 10

HOW AI IS BUILT 12

HOW AI UNDERSTANDS LANGUAGE 15

HOW AI PREDICTS BEHAVIOR 20

HOW AI THINKS (AND HOW IT DOESN'T) 21

PART II: WHAT CAN GO WRONG 23

HOW AI HALLUCINATES 23

HOW WE DECIDE TO TRUST AI 24

PART III: HOW AI ADAPTS TO CLINICAL WORK 27

HOW AI PICKS UP WHAT YOU'RE NOT SAYING 28

HOW FINE-TUNING MAKES AI CLINICALLY USEFUL 29

HOW AI PERSONALIZES CARE 31

PART IV: WHAT'S COMING NEXT IN AI 33

HOW AI BECOMES MORE HOLISTIC 33

WHEN AI BECOMES THE SUBJECT 35

SIMULATED CLIENTS AND DIGITAL TWINS 36

CONCLUSION 38

CHAPTER 3: THE ROLE OF DATA 41

WHAT TYPES OF DATA POWER AI IN MENTAL HEALTH 42

STRUCTURED VS. UNSTRUCTURED DATA 45

DATA QUALITY 47

BIAS IN AI 50

DATA PRIVACY AND CONSENT 53

CONCLUSION 56

CHAPTER 4: AI IN CLINICAL AND COUNSELING PRACTICE 59

ADDRESSING LIMITED ACCESS TO MENTAL HEALTH SERVICES 60

THERAPEUTIC ENGAGEMENT & CONTINUITY 63

AI IN TELEHEALTH AND REMOTE PSYCHOLOGICAL PRACTICE 67

IMPROVING DIAGNOSTIC PRECISION 71

SUPPORTING EARLY DETECTION OF EMERGING MENTAL HEALTH CONCERNS 75

CRISIS MONITORING & RISK DETECTION 79

AI IN CHILD & ADOLESCENT PSYCHOLOGY 83

REDUCING BIAS AND VARIABILITY IN PSYCHOLOGICAL ASSESSMENT 88

IMPROVING TREATMENT SELECTION AND PLANNING WITH AI 92

REDUCING ADMINISTRATIVE BURDEN IN MENTAL HEALTH PRACTICE 96

CONCLUSION 101

CHAPTER 5: AI IN FORENSIC PRACTICE 103

EVALUATING COMPETENCY TO STAND TRIAL 104

DETECTING DECEPTION AND MALINGERING 108

DIGITAL FORENSIC PROFILING 113

SENTENCING AND PAROLE DECISION SUPPORT 117

CROSS-CULTURAL CONSIDERATIONS IN AI FOR FORENSIC PSYCHOLOGY 122

CONCLUSION 126

CHAPTER 6: AI IN WORKPLACE PSYCHOLOGY AND ORGANIZATIONAL WELLBEING 127

AI IN EMPLOYEE RECRUITMENT & SELECTION 128

AI IN EMPLOYEE TRAINING & SKILL DEVELOPMENT 134

AI-POWERED WORKPLACE ANALYTICS & PERFORMANCE EVALUATION 138

AI IN EMPLOYEE WELL-BEING & MENTAL HEALTH MONITORING 143

AI-POWERED DECISION SUPPORT FOR LEADERSHIP & ORGANIZATIONAL STRATEGY 147

CONCLUSION 152

CHAPTER 7: AI IN SCHOOL & EDUCATIONAL MENTAL HEALTH PRACTICE 153

PERSONALIZED LEARNING & STUDENT PERFORMANCE PREDICTION 154

AI IN SPECIAL EDUCATION & INCLUSIVITY 158

ADDRESSING EDUCATIONAL INEQUALITY 162

SUPPORTING STUDENT EMOTIONAL WELL-BEING 165

BOOSTING STUDENT ENGAGEMENT 169

AUTOMATING ADMINISTRATIVE TASKS FOR EDUCATORS 173

CONCLUSION 178

CHAPTER 8: AI IN HEALTH PSYCHOLOGY 179

ENHANCING PERSONALIZATION IN HEALTH PSYCHOLOGY 180

AI IN NEUROPSYCHOLOGY 186

AI IN PSYCHOPHARMACOLOGY & TREATMENT PLANNING 190

AI FOR REMOTE MENTAL HEALTH MONITORING 196

AI-ASSISTED HEALTH BEHAVIOR CHANGE PROGRAMS 201

CONCLUSION 205

CHAPTER 9: AI IN RESEARCH & EXPERIMENTAL PSYCHOLOGY **207**

ENHANCING DATA PROCESSING FOR PSYCHOLOGICAL STUDIES WITH AI 208

AUTOMATING PSYCHOLOGICAL HYPOTHESIS GENERATION 212

IMPROVING EXPERIMENTAL DESIGN AND DATA COLLECTION WITH AI..217

AI AS A REPLACEMENT FOR HUMAN RESEARCH PARTICIPANTS IN PSYCHOLOGICAL STUDIES 221

PREDICTIVE MODELING FOR PSYCHOLOGICAL TRENDS 226

CONCLUSION 231

CHAPTER 10: NAVIGATING REGULATION, PRIVACY, BIAS, AND ETHICS **233**

REGULATION AND GOVERNANCE 234

PRIVACY AND SECURITY 241

BIAS AND FAIRNESS 250

ETHICAL RISKS AND PROFESSIONAL RESPONSIBILITIES 256

CONCLUSION 261

CHAPTER 11: FRAMEWORKS FOR EVALUATING AI TOOLS IN PRACTICE **263**

CLINICAL AND MENTAL HEALTH EVALUATION FRAMEWORKS 264

ORGANIZATIONAL READINESS AND CAPABILITY MODELS 272

ETHICAL RISK & GOVERNANCE FRAMEWORKS...280

EDUCATION AND HUMAN-AI INTERACTION FRAMEWORKS 286

CONCLUSION 293

CHAPTER 12: WHAT'S NEXT **297**

IN-SESSION AI CO-PILOT AGENTS..298

AI WORKFLOW ASSISTANTS 300

BETWEEN-SESSION AI THERAPY ASSISTANTS 302

AI RELATIONSHIP CARE PLATFORMS..305

CLIENT AND PROTOCOL AGENTS 309

CONCLUSION 312

CHAPTER 13: A FEW LAST COMMENTS **313**

TEN CORE BENEFITS FOR AI IN PSYCHOLOGICAL PRACTICE 314

FIVE GUIDING PRINCIPLES FOR RESPONSIBLE AI USE 316

CALL TO ACTION 317

REFERENCES.. **319**

GLOSSARY OF TERMS **338**

Foreword

This foreword was written from the perspective of imagining dialogues with mental health professional colleagues about the core issues, challenges, and opportunities that Artificial Intelligence (AI) presents to the world of mental health practice.

This book invites mental health professionals to move beyond apprehension and toward informed engagement with artificial intelligence, recognizing the unique opportunity our field has to influence how technology is applied in mental health care. Historically, mental health practitioners have often remained on the sidelines during major shifts in health care policy, managed care, and technological innovation, guided by caution and reluctance to act prematurely. Recall that, even just five years ago, very few therapists practiced therapy via online video formats. The pandemic made it a necessity, and soon after it became clear that it had many advantages. Now, it is commonplace.

Will we watch the AI train speed past, or step up as leaders to guide its course?

Few of us will ever dive deep into the technical aspects of AI, and that's okay. What matters is gaining a clear, practical sense of how to use it wisely—and this book delivers just that. With clarity and vision, the authors provide the tools to question, understand, and confidently bring AI into our work in ways that truly guide people across multiple specializations.

It is Here
Artificial intelligence is no longer a futuristic concept. It is a present reality transforming nearly every industry, including psychology—from research to education to the very way we connect with clients. AI is opening new doors for innovation while also forcing us to confront profound ethical questions about how to use it wisely.

AI has already begun influencing mental health practice and research through a variety of tools, including automated assessments, virtual therapy and coaching, interactive chatbots, charting and report software, and large-scale data and qualitative thematic analysis. AI's predictive capabilities offer the potential for early intervention and for identifying at-risk individuals before conditions escalate.

Together, these breakthroughs can make mental health care more accessible, more accurate, and more responsive than ever before.

Engaging and Collaborating with AI

Mental health professionals cannot afford to sit back and let AI happen to them. We need to step in as active, informed guides who know what this technology can and cannot do. Developing AI literacy is now a professional responsibility, much as it was with earlier technological shifts from radio and television to computers, email, the internet, smartphones, and social media. AI is the latest technology, and there will be more to follow—some of which only science fiction has dreamed of. Indeed, science fiction is often, though not always, a good predictor of the future.

As AI rapidly evolves, we must take an active and meaningful role in shaping its development. This means integrating psychological expertise into AI designs and establishing policies that protect privacy and promote outcomes research. For example, how do AI tools affect key aspects of emotional and mental well-being such as symptom relief, coping, resilience, and self-transformation? How can AI tools address the needs of diverse groups—children, adults, seniors, couples, and families—while remaining culturally sensitive? By pursuing these efforts, we aim to ensure that AI develops responsibly and benefits clients, practitioners, and communities.

Mental health professionals can also benefit from developing a working knowledge of AI methods such as machine learning, natural language processing, and decision-support systems to support ethical and effective implementation. We may not write the code, but we can still influence how AI serves clients and shapes the future of mental health care.

Supercharging Self-Help Resources with Interactive AI Tailored to Individuals

Self-help books and online resources already offer practical and accessible tools and skills to millions of people. AI is supercharging these mental health resources through emerging technologies, including AI-guided self-help applications, digital coaching platforms, chatbots, and avatar-based therapy. While most change and healing begin with everyday practices—reading books, talking with friends, self-reflection, spiritual practices and guidance, podcasts, webinars, and chatbots— psychotherapy becomes essential when these practices are not enough.

I am not one to believe AI will replace psychotherapists, but it may revolutionize how we approach self-help. The real promise lies in self-driven tools that transform self-help from something static—like a book, article, audio, or video—into something alive and interactive. Instead of passively consuming advice, people can now engage in real-time dialogue tailored to their unique questions and struggles. That is not just an upgrade; it is a leap forward in making mental health support more accessible, personal, and empowering. Fears of AI decreasing the need or motivation to seek psychotherapy are understandable. However, AI might actually increase the demand for psychotherapy by providing introductory exposure to psychological concepts and motivating individuals to pursue more comprehensive therapeutic engagement.

Ethical and Value Challenges

AI can do amazing things in mental health, but it also raises tough ethical concerns we cannot ignore. If models are biased, if data is not private, or if algorithms operate in secrecy, then these tools have the potential to cause more harm than good. AI tools, if improperly designed, could reinforce disparities in psychological diagnoses and treatment recommendations. Imagine an AI therapist that becomes just a cost-cutting shortcut for insurance companies, offered only to people who cannot afford to see a human therapist.

We must push for AI that is transparent, fair, and accountable, so it truly becomes a partner in healing rather than another barrier to care. AI advancements must be implemented with ethical, values-driven oversight that respects individual differences and cultural contexts. By doing so, we can ensure that AI remains a tool for enhancing psychological care while preserving the human connection at the heart of effective therapy. AI can serve as a powerful ally, enhancing care rather than diminishing it.

AI Therapist or Human Therapist: Which Is Better?

Will AI therapists be even better than human therapists? That is like asking, "What's the best kind of exercise?" Most people know the answer: the best exercise is the one you will actually do and that does not cause injury. Therapy works the same way. CBT, DBT, psychodynamic therapy, self-help resources, AI-based interventions—none is universally better than the others. What matters most is finding the approach that fits the person and the unique situation. Without that context, the question of "better" doesn't make sense.

The so-called Dodo bird verdict—a metaphor drawn from Lewis Carroll's Alice in Wonderland—suggests that all empirically supported psychotherapies are equally effective, largely due to common factors such as the therapeutic alliance. This perspective, however, has been challenged by researchers seeking to identify which techniques work best for specific diagnoses, populations, and individuals. In practice, outcomes depend on multiple variables: the alliance, the client's goals, condition, motivation, learning style, and cultural context, as well as the quality and appropriateness of the interventions. The same principle applies when comparing human-delivered therapy with AI-based interventions. How these approaches will ultimately align and diverge remains an open question—one that calls for continuous investigation.

An esteemed colleague of mine believes that AI can already evaluate and respond like a human therapist, picking up on tone, body language, and even subtle cues. Maybe so—but people relate to each other in ways that are messy, intuitive, and beautifully unpredictable. We react not just with logic, but with instinct, unconscious hunches, and emotion that no algorithm can capture. In 1670, Blaise Pascal wrote, "The heart has its reasons that reason does not know." Our emotional compass and even unpredictability are part of our humanity and wisdom. This is not to dismiss AI or logic, but rather to emphasize that human vulnerability, emotion, and intuition remain strengths that technology cannot fully replicate.

Therapeutic Alliance?
Right now, mental health AI tools work mainly through programmed language responses. They can offer a wealth of information and guidance, and that has real value. But outcomes research reminds us of a deeper truth: what heals in therapy is not just information and knowledge, but the bond of trust between client and therapist, strengthened by the right methods and the client's willingness to grow. When there is trust, therapy works better! A strong therapeutic alliance allows clients to more fully embrace the tools and techniques of therapy and carry them into their daily lives. Similarly, although it may seem counterintuitive, individuals often place trust in authors or online speakers they have never met because they value their words and stories. By the same mechanism, clients may extend trust to AI-based resources, provided those tools deliver valuable feedback and guidance.

Does it matter if words expressing empathy—such as "That must have really been painful" or "I see what you are saying"—come from a therapist or from an AI avatar therapist? At first glance, I would say I would rather hear it from a person. But it is not that simple. People project meaning and trust onto all kinds of figures, whether it is a therapist, a priest behind a screen, an author of a book, a songwriter, or even an AI avatar therapist. While some prefer the anonymity of AI interactions, others value human empathy more.

Most of my clients are using AI in one form or another, including interactive chatbots that offer advice about their lives. Instead of avoiding the topic, I ask them about their use of AI and bring their experiences into the session, just as we do with everything else in their lives. These dialogues have brought up meaningful topics that lead to valuable insights and therapeutic changes, as well as discussions about the limitations of AI resources.

The Future

While some brilliant futurists differ on this matter, AI is not a substitute for human expertise or lived experience. Rather, it is a collaborator, augmenting the skills of trained professionals and expanding the reach of self-help and mental health care.

AI (Artificial Intelligence) is not HI (Human Intelligence), and it is not EQ (Emotional Intelligence). AI does not embody qualities such as love, compassion, creativity, spirituality, intention, or biology—however much it may simulate them through expert programming. Predictions of AI becoming sentient beings have not yet materialized, and if they do, it will constitute a fundamentally new mode of existence.

The future of psychology, with humility in that the future is never fully known, does not have to be AI versus human expertise. Indeed, HI + EQ + AI is the hybrid formula that can open new vistas for positive changes and healing. This adds up to a panoply of opportunities to engage human experience, suffering, hope, and growth in new and meaningful ways.

The cover image of this book says it all: the techno hand and the human hand reaching out to each other in partnership.

Randall Wyatt, PhD
Licensed Psychologist, Psychotherapist
Professor of Clinical Psychology, Director of Professional Training
California School of Professional Psychology, San Francisco Bay
Alliant International University

Preface

When I first began hearing about artificial intelligence (AI) in mental health care, my reaction was probably much like yours: a mix of curiosity and hesitation. AI seemed to be everywhere. In 2025, I counted six Super Bowl ads for products "powered by AI." Closer to my own field, colleagues were discussing AI chatbots that claimed to detect depression. I found myself asking: What exactly is AI? How accurate is it? And how do I know when it is appropriate to use in my profession?

What became clear is that AI is no longer a distant innovation confined to science fiction. For those of us in mental health, it is already here. AI appears in tools that help screen for symptoms before a first appointment, in systems that suggest clinical language for documentation, and in software that analyzes vocal tone for emotional cues. As clinicians, we are being asked to engage with technologies we were never trained to evaluate. This book was written to bridge that gap.

Artificial Intelligence for Mental Health Professionals is for mental health practitioners who want a clear, accessible, and ethically grounded introduction to AI. It is not about hype, advertising, or technical jargon. It is about providing a foundation for making informed, thoughtful decisions about how AI intersects with your clients, your values, and your work.

This book will not turn you into a data scientist—that is not the goal. What it offers is working knowledge: enough to ask the right questions, recognize red flags, and identify when AI can enhance care and when it should be avoided. When AI shows up in your clinical tools or in your clients' lives, you will be able to meet it with confidence and clarity.

To write this, I partnered with two technologists, Vince and Antonio Diego Vasquez, a father–son team developing AI tools to help men choose and become better partners in their intimate relationships. We share the conviction that mental health professionals should not stand on the sidelines. We must take an active role in shaping how AI is used in our field. That begins with understanding what AI is, where it is already being applied, and how to ensure it aligns with the ethics and human connection at the heart of psychological care.

Brenda Hart, Ph.D.
Licensed Psychologist and
Co-Founder of Relationship Workout, LLC
at RelationshipWorkout.com and RelationshipFitness.com

Peer Review Acknowledgment

This book was peer reviewed by licensed professionals to ensure clinical accuracy, relevance, and value to the field of psychology and mental health, and to support independent study that may qualify for CE credit in select U.S. states.

We extend our sincere gratitude to the following reviewers for their time, insights, and commitment to advancing ethical and effective use of AI for mental health practice.

Dr. Andrew Bertagnolli, PhD
Clinical Psychologist, Associate Professor & Healthcare Executive
Associate Professor, Alliant International University
Senior Director of Clinical Strategy, Optum Behavioral Health

Alicia Fisher, LMFT
Licensed Marriage and Family Therapist & Continuing Education Advisor
Specialist in Family Systems Therapy and Youth Programs
Advisory Board Consultant, PESI

Lisa Gray, LMFT
Licensed Marriage and Family Therapist & Author
Private Practice, California
Author of *Healthy Conflict, Happy Couple*

Sula Goldenberg, LMFT
Licensed Marriage and Family Therapist & Former Adjunct Professor
Private Practice, California
Former Adjunct Professor, Pepperdine University

Full peer review forms are available upon request for CE documentation. To request a copy, email: ce@relationshipworkout.com.

Note: This peer-reviewed book may qualify for self-study CE credit in states that recognize independent learning from professional publications. For eligibility details and downloadable documentation, visit www.RelationshipWorkout.com/CE.

Chapter 1: Introduction

Why This Book, and Why Now?

Artificial intelligence (AI) is no longer just a topic in tech headlines—it's quietly entering the field of psychological practice. From therapy apps and risk assessment tools to note generators and diagnostic support, AI is starting to show up in day-to-day psychological work; however, most professionals in the field were not trained to evaluate these tools, much less guide their use.

That gap creates real tension. On one side, AI is framed as revolutionary—able to detect symptoms, reduce administrative burden, and broaden access to care. On the other, it feels opaque, overhyped, and potentially harmful if applied without context, oversight, or ethical safeguards. For many in the field, AI seems either too technical to understand, or too risky to trust; however, we are past the point of asking whether AI will impact psychological practice. The real question now is how—that is exactly what this book is here to help you answer.

This book is not a technical deep dive or a speculative look into the far future. It is a grounded, practical primer for psychologists, counselors, educators, researchers, and others working across the broad spectrum of mental health care. It is designed to demystify AI—what it is, how it works, and where it is already appearing in real-world settings. Just as importantly, it offers foundational knowledge for ethical reflection and informed decision-making, because how we engage with AI now may very well shape its future role in our field.

The objective is not to turn mental health professionals into technologists; it is to help build the fluency needed to ask smart questions, weigh real risks, and make choices that protect the values at the core of our work. AI is not ready to lead; however, it is here, and mental health professionals have a role to play in what happens next.

Who This Book Is For

This book is for anyone working in or preparing for a career in psychological practice—whether you are a clinician, counselor, educator, researcher, or student. If you have started hearing about AI but aren't sure how it fits into your work, this book is for you. It is designed to help mental health and psychology professionals at all levels build a grounded understanding of what AI is, where it is showing up, and how to engage with it thoughtfully, ethically, and with confidence.

It's for the practitioner wondering whether AI will truly reshape the field—or whether it is just another wave of tech-world hype. It is for those who feel overwhelmed by the buzzwords and want a plain-language introduction that answers, "What is this technology? Is it any good? What does it mean for how I deliver care?"

If you are already using AI-powered tools in your work—or guiding others who are—this book can help you ask the right questions to ensure those tools are used responsibly. That includes recognizing when to be concerned about bias, client data security, or regulatory requirements, and understanding AI's limitations to support more informed, ethical decisions.

This is not a guide for computer scientists or engineers. You will not find coding tutorials or algorithm design walkthroughs. Instead, this book offers a plain-language primer that connects the fundamentals of AI with the values, roles, and responsibilities of psychological work.

Wherever you fall on the spectrum—curious, skeptical, unsure, or cautiously optimistic—this book will help you start asking better questions about AI: what it can and cannot do, and what you need to watch out for. Most importantly, it is designed to help you stay grounded in your clinical or professional judgment, while engaging thoughtfully with tools that are becoming increasingly present in the work we do.

How This Book Is Structured

This book is designed as a practical first stop for understanding AI in the context of psychological practice. It is not a technical manual, and it does not assume any background in data science or programming. Instead, it is written for mental health students and professionals who are beginning to encounter AI in their work and want a clearer view of how it functions, where it is being used, and what it means for ethical, effective care.

Chapters 2 and 3 offer a plain-language foundation, explaining how AI "learns" and why its relationship to data is so important. These chapters are tailored for professionals whose expertise lies in helping people—not in writing code.

Chapters 4 through 9 offer a wide-angle look at how AI is already being applied across seven key mental health roles: counseling, clinical, forensic, health, educational, industrial-organizational, and research. Each chapter highlights examples of real or emerging uses of AI—not to be exhaustive, but to show the range of ways AI is already showing up in practice. Some tools are experimental, others are in production, but all demonstrate that AI in psychology is not just theoretical—it is happening now.

Chapters 10 and 11 shift the focus from application to oversight, helping psychologists critically assess the risks and readiness of AI tools. Chapter 10 explores core issues like privacy, bias, and legal risk, while Chapter 11 introduces practical frameworks for evaluating AI in mental health, organizational, and educational contexts. Together, they provide tools for navigating ethical and evidence-based adoption.

Chapter 12 looks ahead. This chapter offers five near-term developments worth watching. It does not try to predict the distant future, but rather helps you anticipate how AI may evolve in ways that could shape your work in the next few years.

Finally, Chapter 13 concludes the book by summarizing the benefits of AI in psychological practice and outlining guiding principles for its responsible use.

Together, these sections strive to provide a practical, ethical, and profession-centered foundation for engaging with AI—not as something to fear or blindly adopt, but as a set of tools that require your expertise to be used wisely.

How to Use This Book

There is no one right way to read this book, as it is designed to be flexible depending on your interests and needs.

One option is to read it start to finish or at least begin with the early chapters, which lay the groundwork for understanding AI in plain terms, while later chapters build on that foundation with discipline-specific examples, ethical considerations, and a forward look at what is coming next.

Another approach is to begin with the chapter that aligns most closely with your professional role—clinical, counseling, forensic, health, educational, industrial-organizational, or research. Within each of these chapters, you will find five to ten examples on how AI is being applied in that specific field of mental health. From there, you can return to the earlier chapters 2 and 3 to better understand the foundational AI concepts behind the tools you are most curious about.

Wherever you start, this book is meant to be a practical companion. You do not need to be a technologist to understand it. You just need your existing clinical instincts, ethical judgment, and a healthy curiosity about the tools reshaping our field.

Disclaimer on Company and Product Mentions

Throughout this book, we reference a range of companies, tools, and platforms to illustrate how AI is being researched and applied in mental health. These mentions are for educational purposes only and do not constitute endorsements or recommendations. Inclusion does not imply validation of a product's efficacy, quality, or market leadership.

Where specific tools are mentioned, they were selected based on one or more of the following: presence in peer-reviewed literature, relevance to the topic at hand, public visibility, or representation of key trends in the field. Most references are drawn from over 100 peer-reviewed articles and reputable public sources.

At the time of writing and publication, the authors have no financial interest in any company or product mentioned in this book, except for Relationship Workout, LLC, which offers an AI-driven Relationship Care Platform. We reserve the right to invest in or receive compensation from referenced companies or tools in the future.

Given the fast pace of innovation, some tools or companies mentioned may change, merge, or no longer exist by the time of reading. We will make every effort to keep future editions current.

Continuing Education Credits and Content Updates

Artificial intelligence is evolving rapidly, and its applications in psychology continue to grow. To help readers stay informed, we plan to share updates to this content as well as additional educational materials, such as articles and related resources.

This book—and its companion content—may also qualify as foundational material for continuing education (CE) credit, depending on your licensing board and the type of credit you seek.

To receive updates or learn more about CE opportunities, scan the code below or visit RelationshipWorkout.com/CE.

A Final Word Before We Begin

AI is already beginning to shape psychological work, but it may not be always clear how, or what it means for your specific practice. This book does not attempt to offer all the answers. What it does offer is a starting point: a way to start to make sense of the unfamiliar, to ask smart questions, and to begin to engage with AI thoughtfully and ethically. We suggest that you do not need to become a tech expert to understand AI. Rather, you would do well to have just enough clarity to make more informed choices on when AI might help, when it might not, and how to keep the focus where it belongs—on human care.

Chapter 2: Demystifying AI

Artificial intelligence is moving quickly into the world of mental health, but many professionals feel unsure about how to evaluate or use these tools. This chapter is designed to help bridge that gap. You do not need training in computer science to follow along. What you do need is a clear, clinician-friendly foundation.

AI tools are already helping with many clinical tasks, from documentation and assessment to supporting therapy sessions. To use them responsibly, it is important to understand how these tools generate insights, what they can do, and what they cannot. Without this understanding, it is easy to place too much trust in the technology, overlook important risks, or use it in ways that may affect client care.

This chapter introduces the essential building blocks of modern AI, explained with metaphors connected to clinical training. The goal is to help mental health professionals engage with these tools in a way that is both confident and critical, ensuring that AI strengthens rather than undermines ethical, effective psychological care.

The chapter is organized into four parts:

- **Part I: How AI Works.** Explains the basic concepts behind AI, using clinician-friendly metaphors to show how these systems learn, reason, and generate responses.

- **Part II: What Can Go Wrong.** Reviews common limitations, risks, and ethical concerns with AI, such as bias, hallucinations, and overreliance, so clinicians are better informed when using these tools.

- **Part III: How AI Adapts to Clinical Work.** Explores how AI tools can be tailored for specific therapeutic settings, client groups, and professional roles through fine-tuning, personalization, and multimodal inputs.

- **Part IV: What's Coming Next.** Looks ahead to new possibilities such as integrated AI systems, "machine psychology," and simulated clients that may reshape the future of practice.

How AI Learns

❖ **What You'll Learn:** How AI systems learn by being exposed to data and feedback, similar to how clinicians build insight through supervised experience.

Clinicians develop expertise by noticing patterns, such as recognizing client behaviors, interpreting symptoms in context, and adjusting interventions based on outcomes. Artificial intelligence works in a similar way, but instead of learning from lived experience, it learns from data.

You can think of AI as a trainee therapist beginning practicum. It starts with no knowledge or insight, only a blank slate shaped by examples. Where a trainee relies on supervision, AI relies on data. Where a trainee draws on intuition, AI detects statistical patterns. With enough examples, AI begins to make predictions. Some are useful, others less so, depending on the quality and diversity of the data it has received.

To understand what AI can and cannot do in mental health, it is helpful to begin with how it learns.

AI Learns through Examples, Not Instructions

Traditional software runs on fixed instructions, like an actor following a script. It only does what it is told, with no ability to adapt or improvise.

Early versions of AI worked this way too. Engineers created detailed rules for every situation. These systems could seem intelligent, but only within the narrow limits of their programmed rules.

Modern AI is different. It uses a process called machine learning. Instead of being told what to do step by step, the system is trained on examples, called "training data." It does not memorize answers. It learns patterns, then makes predictions based on those patterns.

In short, machine learning allows an AI system to get better the more examples it processes.

> **Machine learning** is a broad field of AI focused on developing systems that can learn from data and make predictions or decisions without being explicitly programmed for every task.

In other words, just as a psychology student learns by working with clients and studying case material, AI learns by being exposed to examples. For instance, to recognize patterns in human language, ChatGPT was trained on hundreds of billions of words, equal to millions of books.

But unlike a student, AI does not understand meaning. It does not think or feel. It identifies patterns in the data it was trained on and uses them to predict what is likely to come next. Its accuracy depends on the quality of those training examples.

Approaches to Machine Learning Training

There is no single best way to train either a student or an AI system. The method depends on how the learner receives information and feedback. In psychology, a trainee may learn by reviewing case files, recognizing behavior patterns, or receiving supervision. AI training follows a similar logic and usually falls into three main types: supervised learning, unsupervised learning, and reinforcement learning.

Supervised Learning

This is like a structured practicum. The AI is trained on data where each example includes both the input and the correct answer. It is similar to reviewing a clinical case where the diagnosis is already known. Over time, the system learns which features point to which outcomes.

This method has been used to create tools that, for example, help identify suicide risk by detecting patterns in patient history (Dwyer, Falkai, & Koutsouleris, 2018). But just like with human trainees, if the examples are too narrow or biased, the AI's conclusions can end up skewed and unreliable.

Unsupervised Learning

This is more like giving a student 500 anonymized case notes—with no diagnoses or labels—and asking them to find patterns on their own. The AI is not given answers during training. Instead, it looks for similarities and groups the data into clusters.

This approach is often used in research, such as identifying subtypes of trauma response or patterns of co-occurring symptoms. Like a student forming hypotheses from observed trends, unsupervised learning can bring out new insights. Yet without structure or validation, those insights may be harder to interpret or use in clinical care.

Reinforcement Learning

This resembles learning through trial and error. The AI receives feedback for its actions, with rewards for helpful responses and penalties for mistakes. Over time, it adjusts to maximize positive outcomes. In clinical terms, it is like how a trainee refines their approach based on client reactions, gradually learning which interventions are most effective (Lin, Cecchi, & Bouneffouf, 2023).

But just like with human learners, reinforcement alone is not enough. Without careful oversight, the AI may focus on the wrong signals, such as valuing surface-level engagement over meaningful therapeutic progress. In some cases, it might technically reach its goal, but in ways that are inefficient or even clinically inappropriate. That is why training objectives must be aligned with outcomes that truly matter in mental health practice.

★ **Key Takeaway:** AI learns by example, not by instruction; the quality and variety of its training data shape how well it performs in the real world.

How AI Generates Answers

❖ **What You'll Learn:** Why modern AI creates answers by making predictions using probabilities generated through its training, and what this means for interpreting its output.

Early in training, psychology students often want clear rules: if a client says one thing, respond in a set way. With experience, they discover that therapy is not black and white. Good decisions depend on context, lived experience, and clinical judgment about what is most likely to help.

AI has followed a similar path.

The earliest AI systems were rule-based. Programmers wrote decision trees: if X happens, then do Y. These systems could seem intelligent, but only within narrow, pre-defined boundaries. They had difficulty when faced with situations that did not fit their rules.

Modern AI works differently. Instead of strict instructions, it uses probabilistic reasoning. It scans thousands or even millions of examples in its training data and estimates what is most likely to come next. This is less like following a script and more like recognizing patterns on a massive scale.

> **Probabilistic reasoning** is how AI makes predictions: by calculating which response is most likely based on patterns it has seen before. It's a statistical process—not intuition or understanding—that helps the AI estimate what's likely to come next based on probabilities learned during training.

To see this in practice, suppose you type, "Write a four-line poem about anxiety" into an AI tool.

A human might think about tone, meaning, and metaphor. AI does not. It breaks the input into small pieces called *tokens*, usually words or parts of words. Then it builds its response one token at a time, predicting what is most likely to come next based on patterns from its training data.

> **Tokenization** is how AI breaks down text into smaller units—called tokens—so it can predict one word (or word part) at a time based on statistical likelihood.

Let's say the model has already generated the phrase: "The sky was dark and full of." Now it must predict the next word. Based on patterns in its training data, it might select "stars." It does not understand poetry or emotion. It simply recognizes that "stars" often follows that phrase in similar contexts. The process repeats one token at a time until the four-line poem is complete.

The poem may sound poetic or thoughtful, but the AI is not being creative or insightful. It is filling in blanks based on patterns it has seen before.

This explains why AI systems can sound confident even when they are wrong. They are not reasoning or reflecting. They are making statistical predictions. The answer may be the most likely one, but that does not guarantee it fits the situation. It is like assuming a therapy will always work just because it helped other clients with similar symptoms. Every case is different, and context matters.

For clinicians, the takeaway is this: even when AI responds with certainty, treat it as a second opinion, not a final answer. These tools cannot replace clinical expertise, emotional understanding, or ethical judgment. They can suggest possibilities, but they cannot determine what is right for the individual in front of you.

★ **Key Takeaway:** AI doesn't "know" the answer; it calculates what's most likely to come next based on prior patterns, making it powerful but not always right.

How AI is Built

❖ **What You'll Learn:** The basics of neural networks and deep learning, and why the complexity of these models makes their reasoning difficult to explain.

So far, we've seen that machine learning allows AI systems to learn from examples rather than following a fixed set of rules. Within this approach, two methods are especially important for understanding complex, real-world situations in mental health care: neural networks and deep learning.

- **Neural networks** are loosely based on how the brain processes information. They learn by layering information, which allows the AI to gradually detect patterns. For example, they can pick up on how certain words, tones, or behaviors tend to appear together.

- **Deep learning** is a more advanced form of neural networks. It uses many additional layers, which allows it to notice subtle or abstract patterns. This is especially useful for large or messy data sources such as therapy notes or voice recordings.

These methods, especially deep learning, power the tools behind ChatGPT and Gemini. Their arrival marked a turning point. What once seemed abstract or futuristic became real: tools that could respond in fluent language, analyze text, and even simulate aspects of dialogue. For many clinicians, this was when AI shifted from research conversations into everyday conversation and practice.

Understanding how these systems are built helps clinicians recognize both their potential and their limitations.

Neural Networks

Neural networks process information step by step, with each stage noticing something new. You can imagine it like a team of people, each with a unique skill, examining the same information in turn. One person might notice a simple detail, the next might see a deeper pattern, and by the end, the team has built a fuller picture. This teamwork allows the system to detect patterns that aren't obvious at first glance.

A neural network usually has three main parts:

- **Input layer**: This is where information enters, such as speech, text, or behavioral data.

- **Hidden layers**: These sit in the middle and do most of the work. Each one passes the information through filters, adjusting the way the system interprets it at every step.

- **Output layer**: This provides the final result, such as predicting an emotion, assigning a risk score, or labeling data as "depressed" or "not depressed."

The hidden layers are where the real power lies. As information moves through them, the system detects complex relationships that would be impossible to hand-code. This is what makes neural networks so effective. But because the learning happens across so many layers, it isn't always clear how the final answer was reached. That lack of transparency becomes even more challenging in advanced systems, which leads us into deep learning.

Deep Learning

Deep learning is a more advanced form of AI that builds on neural networks by adding many extra layers of processing. This added complexity makes the systems more powerful, but it also makes it harder to understand how they reach their decisions. This issue is often called the "black box" problem. It is especially important in clinical settings, where openness and accountability are essential.

> The **"black box" problem** refers to how difficult it is to understand or explain exactly how an AI system—especially deep learning models—arrived at a specific decision. The system may give an answer, but the reasoning behind it is hidden, making it hard to verify, question, or fully trust in clinical settings.

Think of deep learning as the difference between a beginner trainee and someone with years of clinical experience. With more exposure, the seasoned trainee notices subtler, more layered patterns in how people speak, act, or respond.

Systems like ChatGPT and Gemini brought this approach into the public spotlight. These models were trained on enormous amounts of text, conversations, and other data—over a trillion data points in the case of GPT-4. As they process this material, they continually adjust their internal settings, refining their predictions much like instincts sharpen with repeated experience.

In mental health, deep learning makes it possible for AI tools to:

- Notice early signs of cognitive decline based on language shifts.
- Flag relapse risk by detecting changes in tone or emotion across therapy sessions.
- Detect patterns that even experienced clinicians might overlook, provided similar examples were included in training.

These capabilities are powerful, especially when applied at scale, but they come with important limits. While the system may recognize patterns, it does not truly understand them. Just as a trainee might misinterpret a case without realizing it, an AI model can spot a signal but misread its meaning.

Unlike a clinician, who can pause, reconsider, and explore multiple perspectives before responding, a deep learning system generates its answer in one step. It does not reflect, ask clarifying questions, weigh context, or account for cultural nuance.

And it will not revise its own response unless prompted. For these reasons, human oversight remains essential.

★ **Key Takeaway:** The complexity that gives AI its power also makes it difficult to understand how it reaches its conclusions. For this reason, clinicians must think critically about how they use these tools and apply their suggestions carefully.

How AI Understands Language

❖ **What You'll Learn:** An introduction to how AI approaches human language, including the core techniques it uses to generate text, analyze tone, and detect patterns. This section outlines the building blocks behind tools like ChatGPT and explores their potential role—and limitations—in clinical settings.

For mental health professionals, language is one of the most important tools for understanding and supporting clients. Whether it is listening for meaning in therapy, coding responses in assessments, or documenting progress, language is central to the work. For AI to be useful in this space, it must be able to work with language in flexible and meaningful ways.

This is where technologies such as Large Language Models (LLMs), Generative AI, Natural Language Processing (NLP), and Sentiment Analysis become relevant.

To return to the training metaphor, this is like the stage where a student begins sitting with real clients. They are learning to listen carefully, communicate clearly, and pay attention not only to what is said, but also to what is meant. AI cannot truly understand in the way people do, but it can begin to approximate this process, generating responses that feel more fluent, even human, though without any internal awareness.

Large Language Models

Large Language Models (LLMs), such as those used in GPT-4 and Claude, are a type of deep learning system designed to work with language. These models can carry out many different language-based tasks. They can generate fluent narratives, summarize clinical notes, simulate dialogue, and draft practice-related content.

> **Large Language Models (LLMs)** are AI systems trained on massive text datasets that generate and manipulate human language, powering tools like ChatGPT and Claude.

Large Language Models (LLMs) are trained on very large collections of text gathered from publicly available sources like books, articles, forums, and websites. Most of this material comes from the public domain or open licenses, but some models may also have included copyrighted content, which continues to raise legal and ethical concerns.

In clinical practice, LLMs can be used in several ways, such as:

- Simulating therapy sessions for training or supervision
- Drafting progress notes from transcripts or structured inputs
- Producing psychoeducational content tailored to specific diagnoses
- Offering alternative ways of framing client narratives or case formulations

Clinically, an LLM is somewhat like a very verbal trainee who has read widely but has no lived clinical experience. These systems can create polished and fluent text, but they lack context, diagnostic reasoning, and cultural sensitivity. Without supervision, their work may sound accurate yet contain information that is clinically inappropriate or ethically concerning.

It is also important to note that LLMs do not check facts. As discussed earlier, they generate responses by predicting patterns in their training data rather than pulling from verified sources. This means they can produce fabricated citations, use diagnostic terms incorrectly, or state inaccuracies with confidence. For clinicians, this highlights the need for careful judgment and ethical oversight when using AI-generated text in practice.

★ **Key Takeaway:** Large Language Models generate fluent responses by identifying statistical patterns in massive text datasets, not by understanding meaning. Their outputs can sound human-like, but they lack insight, self-awareness, and contextual judgment, which makes human oversight essential in clinical and ethical settings.

Generative AI

While Large Language Models (LLMs) focus on language, they are part of a broader group of tools known as generative AI. These systems are designed to create new content, such as text, images, audio, or video. Some can generate spoken responses that sound like a real person. Others can draw illustrations, compose music, or create short videos. At their core, they all use the same principle: recognizing patterns in data and generating new material that fits those patterns.

In mental health contexts, these non-text forms of generative AI are becoming increasingly relevant. For example:

- **Visual metaphors:** Image generators can help clients express emotions by creating symbolic representations of feelings like anxiety, depression, or resilience.
- **Simulated voices:** Voice synthesis can support clinician training by simulating clients with different emotional tones, offering practice for challenging conversations.
- **Psychoeducational video:** AI-generated videos can illustrate therapeutic concepts such as grounding techniques, communication strategies, or emotional regulation.

These tools expand the ways clinicians can teach, train, and support clients. However, just like LLMs, they do not truly understand what they produce. Their outputs must be carefully reviewed for clinical relevance, accessibility, and ethical alignment before being used in practice.

Natural Language Processing

Natural Language Processing (NLP) is the branch of AI that allows machines to work with human language in a structured way. It breaks language into parts such as syntax, meaning, tone, and key themes. By doing this, AI systems can analyze, classify, and generate text in ways that are useful for clinical and educational applications.

> **Natural Language Processing (NLP)** is a branch of AI that helps machines make sense of human language. It can analyze, organize, and pull useful information from transcripts, notes, or even spoken words, making it easier to work with large amounts of client or clinical data.

NLP is the foundation of many AI tools used in clinical work. It powers automatic transcription of therapy sessions, extracts symptoms from intake forms, generates alerts from language cues, and detects emotion or tone in client speech or writing. In practice, NLP helps organize unstructured information, such as progress notes or transcripts, into formats that can support documentation, supervision, or outcomes tracking.

While NLP does not understand language the way humans do, it can detect linguistic patterns linked to psychological states. For instance, it might flag disorganized speech in a mental status exam, notice subtle vocabulary shifts that suggest mood changes, or highlight repeated use of cognitive distortions in session transcripts.

However, NLP also has limits. It can misread metaphor, sarcasm, or idiomatic expressions. A dark joke could be flagged as a safety risk, while emotionally flat language might be overlooked. NLP finds patterns, not meaning, and this distinction is critical in clinical care.

> ★ **Key Takeaway:** NLP tools can highlight language patterns that may support clinical insight, but they do not grasp meaning. They may misinterpret sarcasm, miss emotional nuance, or trigger false alarms.

Sentiment Analysis

Sentiment analysis is a type of natural language processing that estimates emotional tone. It often categorizes language as positive, negative, or neutral, and in some cases can detect more specific states like sadness, anxiety, or anger.

In mental health settings, sentiment analysis can be used to:

- Track emotional changes across therapy sessions
- Flag distress signals in client journals or self-guided exercises
- Monitor tone in digital communication, such as texts or therapy apps

Sentiment Analysis is an AI technique that detects emotional tone in language—such as positivity, negativity, or distress—based on word choice, phrasing, and context.

This process is like a practicum student trying to read the emotional undertones in a session transcript. They may notice words that suggest withdrawal, hostility, or anxiety, but without nonverbal cues, pacing, or context, their interpretation may be incomplete. Sentiment analysis systems work the same way. They rely only on text, which makes them useful but also limited.

As with all AI tools, sentiment analysis should be treated as a starting point, not a final judgment. It can highlight potential shifts or warning signs, but only a clinician can interpret those signals within the broader therapeutic context.

★ **Key Takeaway:** AI systems can generate and analyze language with impressive fluency, but they do not grasp meaning, context, or nuance in the way clinicians do. Whether used for drafting text, analyzing tone, or flagging potential risks, these tools are only helpful when guided by clinical reasoning, cultural awareness, and ethical oversight.

How AI Predicts Behavior

As psychology trainees gain experience, they begin forming hypotheses such as, "Based on this client's history and presentation, what outcomes are likely?" This ability to anticipate and intervene is a central part of clinical reasoning. AI systems attempt something similar through what is called predictive analytics.

Predictive analytics uses statistical patterns in past data to estimate the chances of future events. In mental health, this might involve forecasting the risk of relapse, predicting dropout, or identifying which clients may respond best to a specific intervention. These insights are not based on intuition, but on calculations across large sets of prior cases.

> **Predictive analytics** refers to using patterns from past data to estimate future outcomes. In psychology, it can help assess risks such as relapse, treatment dropout, or how likely a client is to respond to a particular intervention. Unlike intuition, these insights come from analyzing data across many prior cases.

Just as clinical judgment improves with exposure to many cases and good supervision, AI predictions depend on the quality of the data used to train them. If that data is incomplete, narrow, or biased, the predictions may carry the same limitations.

It is also important to remember that predictive models are probabilistic, not certain. They do not give definitive answers, but rather estimate likelihoods—for example, suggesting a 70% chance that a client may drop out of treatment. For clinicians, this distinction is critical. Predictive tools can provide useful input, but they cannot capture individual context, emotional nuance, or cultural meaning. They provide insight, not understanding.

When used carefully, AI predictions can serve as decision support, similar to a second opinion informed by data. But they should never replace clinical judgment. As with any consultation, their outputs require professional scrutiny before being applied to client care.

★ **Key Takeaway:** Predictive analytics can reveal patterns and estimate risks, but it does not understand people. These tools provide data-driven input, not clinical wisdom, and must always be interpreted through human judgment, context, and ethical care.

How AI Thinks (and How It Doesn't)

❖ **What You'll Learn:** Why most AI tools used in psychology today are considered narrow, how they differ from hypothetical general AI, and why this distinction matters for selecting the right tools and using them responsibly in clinical care.

Imagine a psychology student early in training. They may be able to conduct an intake or recognize cognitive distortions, but they would likely struggle with complex decisions or culturally sensitive cases. Their skills are real but limited to what they have practiced.

That's how today's AI tools function. They are examples of *narrow AI*, designed for specific tasks such as summarizing text, transcribing speech, or flagging suicide risk. Within those boundaries, they can be highly effective. Outside of them, their limits become obvious.

Narrow AI refers to artificial intelligence systems designed for one specific task, such as generating text, detecting faces, or flagging clinical risk. They perform well within their training boundaries but break down outside them. All AI currently used in mental health is narrow AI.

By contrast, *general AI* refers to a still-hypothetical form of artificial intelligence with broad, human-like reasoning. Such a system could handle uncertainty, move between tasks, and adapt to new clinical situations without needing retraining. It would be the AI version of an experienced clinician—flexible, intuitive, and able to draw on context across different domains.

But general AI does not exist today. It isn't found in clinical tools or in research labs. In fiction, it appears as machines like HAL 9000 in *2001: A Space Odyssey* or the Replicants in *Blade Runner*, which are portrayed as self-aware and capable of genuine insight.

> **General AI** is a still-theoretical type of artificial intelligence that could learn, reason, and carry out any intellectual task a human can. Unlike today's narrow, task-specific systems, a true general AI would be adaptable, autonomous, and capable of making context-sensitive judgments. Importantly, it does not exist yet.

Why This Distinction Matters

Mental health practice rarely fits neatly into categories. A client's language, culture, diagnosis, and lived experience often overlap in subtle and unpredictable ways. An AI model trained to detect depression from vocal tone, for example, might miss similar signs of distress related to grief, PTSD, or culturally specific expressions.

Just as we wouldn't expect a practicum student to complete a forensic evaluation without close supervision, we cannot expect a narrow AI model to perform well outside the area it was trained for. Recognizing these limits helps prevent overreliance and supports ethical, informed use. AI can be helpful, but only if we remember the type of intelligence it was designed to represent.

> ★ **Key Takeaway:** If you plan to use AI in clinical work, confirm that the tool was trained for that specific purpose. Today's AI is narrow. It can be highly capable in some areas but is prone to errors outside its scope. Knowing what a tool was designed to do is essential for using it safely, ethically, and effectively.

———————— ▪●◣ ————————

PART II: WHAT CAN GO WRONG

This section explores the risks, blind spots, and ethical concerns that can arise when AI is used in mental health. It highlights the need for clinical oversight and critical thinking. By the end, you'll be able to identify common failure points like hallucinations, understand why explainability and human oversight matter, and make more informed decisions about when and how to use AI safely in practice.

How AI Hallucinates

❖ **What You'll Learn:** Why AI-generated content can sound confident yet be completely incorrect, how hallucinations occur, what forms they may take in clinical work, and why clinician oversight is essential for accuracy, safety, and ethical care.

Even well-trained models can still make mistakes. One of the most serious risks in clinical use is *hallucination*. This happens when a system produces information that sounds correct but is false, fabricated, or irrelevant.

AI hallucination happens when a system generates responses that sound fluent and confident but include false, fabricated, or misleading information.

Hallucinations in AI can appear in several ways:

- Citations to studies or journals that don't exist
- Clinical terms used inaccurately or out of context
- Confident but inaccurate answers built on flawed or incomplete data

It's like a student confidently citing a study they never actually read because the phrasing feels familiar. In mental health care, this can be risky. A hallucinated diagnosis, misused term, or invented treatment protocol can confuse clinicians, mislead clients, or undermine trust.

These errors are not accidents. They result from how generative AI works. The system predicts the most likely next word in a sequence, rather than checking information against a verified source. It isn't fact-checking—it's assembling text that statistically "fits" the input, even if the output is fabricated.

A helpful way to picture this is to think of autocomplete on your phone. If you type, "I went to the..." and it suggests "store," it's not verifying where you went. It's predicting the most likely completion based on patterns it has seen before. Generative AI does this at scale, producing entire paragraphs that may sound correct but may have no grounding in fact.

That's why clinician oversight is crucial. AI-generated material must always be reviewed through professional judgment, ethical reflection, and cultural awareness. A hallucination is not just a technical quirk—it is a clinical risk that requires careful supervision.

★ **Key Takeaway:** AI-generated language can sound confident and polished even when it is entirely wrong. These errors, called hallucinations, are not software bugs but a natural outcome of how AI produces text. Without careful review, they can create clinical risks. Human supervision is not optional, it is essential.

How We Decide to Trust AI

❖ **What You'll Learn:** How explainability and human oversight help to keep AI tools in mental health transparent, ethical, and clinically useful.

As AI systems become more advanced, it is no longer enough to ask what they can do. We also need to ask whether, and when, we can trust them. In clinical work, trust is not just about accuracy. It also depends on transparency, accountability, and human oversight.

Mental health professionals are trained to explain and defend their decisions to supervisors, clients, and licensing boards. AI should be held to the same standard.

An impressive response means little if it cannot be explained clearly or placed in the context of a client's real experience.

This section introduces two important practices that help make AI safer and more responsible in mental health settings:

- **Explainable AI (XAI)**, which makes a model's reasoning more visible and easier to understand.

- **Human-in-the-Loop AI (HITL)**, which ensures that clinicians stay directly involved in interpreting and guiding AI outputs.

Together, these practices highlight a central principle. AI can support clinical work, but human professionals must always remain in charge.

Explainable AI (XAI)

In clinical training, it is not enough for a psychology student to give the right answer. They must also explain how they reached their conclusion. Was it the client's language, their affect, or a pattern of symptoms? Supervisors expect this kind of transparency because the reasoning is just as important as the result. The same expectation should apply to AI.

Many advanced AI systems, especially those built on deep learning, work like black boxes. They generate outputs such as risk scores or diagnostic suggestions without showing how those conclusions were made. In mental health practice, this is a serious concern. Decisions must be transparent, defensible, and sensitive to culture. Without clear reasoning, AI outputs fall short of these standards.

Explainable AI refers to tools that make it clearer how an AI system reached its conclusions. These tools show which elements, such as word choice, tone, or how often a symptom appears, influenced the result.

Conceptually, explainable AI is like asking a trainee to walk through their case formulation. What information stood out? What was ruled out? How did they connect the dots? This kind of transparency allows psychologists to see whether the AI's suggestions fit with therapeutic goals and their own clinical judgment.

For example, some explainable AI tools can give clinicians real-time feedback on patient language during digital assessments. Instead of only flagging "elevated distress," the system might show that repeated self-blame or hopeless phrases led to the concern. This lets clinicians check whether the AI's reasoning makes sense and whether it supports or challenges their own impressions.

Explainability is more than a technical detail. It is an ethical safeguard. Without it, clinicians risk falling into *automation bias,* where they might accept AI outputs without adequate reflection.

> **Automation Bias** happens when people place too much trust in AI-generated results. This can cause clinicians to overlook mistakes or accept suggestions without enough critical review.

This risk is especially high in sensitive areas like forensic evaluations, involuntary holds, or treatment decisions involving vulnerable populations. Even when a model's reasoning is visible, it still requires human interpretation.

Ultimately, the psychologist—not the model—is responsible for ensuring AI-generated insights are appropriate, evidence-based, and ethically sound.

Human-in-the-Loop AI

Even the most skilled psychology trainee still works under supervision. Oversight is needed to catch mistakes, uphold ethical standards, and bring in professional judgment. AI systems, no matter how advanced, should be used with the same kind of supervision.

> **Human-in-the-Loop AI** is a way of using technology that keeps the clinician at the center of decision-making. The AI can generate suggestions or highlight patterns, but the clinician reviews, interprets, and decides what is useful. This helps to ensure that care remains guided by professional judgment and not by automation alone.

Human-in-the-loop (HITL) systems are designed to keep clinicians involved in how AI is used. These tools do not make decisions on their own. Instead, they depend on trained professionals to judge whether an AI suggestion is valid, safe, and appropriate for the situation.

Think of it like supervision. The AI is the student who offers a hypothesis. The clinician, like the supervisor, reviews that idea in light of the full clinical picture.

In practice, this might look like:

- A therapist checking an AI-generated summary before adding it to the record.

- A forensic evaluator confirming the reasoning behind a risk score before including it in a report.

- A clinician reviewing AI alerts about symptom changes and deciding if or how to respond.

Just as no supervisor signs off on every conclusion from a practicum student without review, clinicians should not accept AI outputs at face value. A human-in-the-loop approach protects accountability, safeguards clients, and ensures decisions remain grounded in professional judgment.

★ **Key Takeaway:** AI can support clinical practice, but only under human guidance. Like any psychological tool, its output needs to be explained, interpreted, and supervised. With thoughtful oversight, even an imperfect AI system can contribute to ethical and evidence-based care.

——————— ⁖ ———————

PART III: HOW AI ADAPTS TO CLINICAL WORK

This section looks at how AI can be adjusted to better fit the needs of clinical practice. AI tools can be tailored for specific groups, personalized for individual clients, and extended to work with new kinds of information. Approaches such as combining multiple data sources, refining a system for a particular task, or drawing on client-specific materials can make AI more relevant, accurate, and supportive of ethical care.

By the end of this section, you'll know how to assess whether an AI tool has been adapted for your therapeutic setting and how to use it in ways that improve accuracy, personalize care, and support client engagement.

How AI Picks Up What You're Not Saying

❖ **What You'll Learn:** How multimodal AI can pick up on cues clients may not voice directly—such as changes in tone, facial expression, or data from wearable devices—and why these signals only gain meaning when interpreted through your clinical judgment and therapeutic relationship.

Multimodal AI systems are designed to capture more than just words. By integrating various input channels—such as vocal tone, facial expressions, body language, and physiological signals—these tools aim to construct a fuller picture of human behavior and emotional state.

> **Multimodal AI systems** are designed to look beyond just words. They bring together different kinds of input, such as vocal tone, facial expression, body language, and even physical signals like heart rate. By combining these sources, the system can create a more complete picture of a person's behavior and emotional state.

In mental health, multimodal AI tools are being developed for several uses:

- **Voice analysis** to pick up cues like anxiety, hopelessness, or agitation from changes in pitch, speed, or tone.

- **Facial recognition** to identify microexpressions that may signal distress, dissociation, or anger, even when words say otherwise.

- **Wearable devices** to track sleep, heart rate, and movement, which can shed light on mood, stress, or relapse risk.

- **Behavioral sensors** to notice shifts in posture, eye contact, or activity over time, especially useful in telehealth or long-term care.

The student metaphor still applies here. Just as a trainee learns to look beyond words—for example, when a client says "I'm fine" but avoids eye contact or appears exhausted—multimodal AI also tries to notice what's left unsaid.

These tools show promise for early detection, continuous monitoring, and adding context in situations where cues are harder to pick up, such as remote therapy. For instance, they might help flag signs of emotional decline between sessions.

At the same time, they can misinterpret signals. A neutral expression may be misread as sadness, or a trauma-related reaction might be flagged as abnormal when it's actually expected. Nuance in sarcasm, avoidance, or masking can also be missed.

The concern isn't only mistakes, but the risk of taking AI outputs at face value without weighing culture, context, or therapeutic rapport.

That's why clinicians must remain the interpreters, not simply the users, of these tools. Multimodal AI can support care, but its insights are only valuable when filtered through clinical judgment, cultural awareness, and the therapeutic relationship.

★ **Key Takeaway:** Multimodal AI can add depth by noticing cues beyond words, but its value depends on the clinician interpreting them. These tools only support care when guided by human judgment, cultural awareness, and therapeutic context.

How Fine-Tuning Makes AI Clinically Useful

❖ **What You'll Learn:** This section explains fine-tuning, a process where large language models are trained further on specialized data. Fine-tuning helps align AI with the language, tasks, and needs of specific clinical settings and populations.

Every psychology student begins with broad training, including introductory courses, theories, and a shared clinical vocabulary. But to be effective, they eventually need to specialize. One student may pursue EMDR to work with trauma, while another may focus on assessments for neurodivergent youth. Their education narrows to meet real clinical needs.

Fine-tuning works the same way for large language models (LLMs). After broad training on general language, they are further trained on specialized data so they can better serve specific contexts, like clinical practice.

> **Fine-tuning large language models (LLMs)** means training them on specialized data so they can better serve a particular purpose. This might involve tailoring a model to a specific population, a clinical task, or a domain such as providing relationship guidance. The goal is to make the system more accurate, appropriate, and useful within that focused context.

General-purpose LLMs are trained on massive collections of open text. They can generate fluent responses, but without additional guidance they may lack the nuance of therapeutic language, clinical reasoning, or cultural awareness. The output may sound polished but still end up vague, generic, or even clinically inappropriate.

Fine-tuning helps to address this gap. It retrains the model on carefully chosen material that matches a specific clinical purpose. In mental health care, this might include:

- Therapy transcripts from multiple modalities
- Evidence-based treatment protocols
- Diagnostic standards from resources like the DSM-5 or ICD-11
- Culturally adapted psychoeducational resources
- Clinical documentation such as progress notes and case summaries

By learning from these targeted examples, the model becomes better able to "speak the language" of the setting where it will be applied. It can draft CBT-style progress notes, generate clear and appropriate psychoeducational handouts, or create intake summaries that follow clinical conventions.

Using the training metaphor: fine-tuning is like when a psychology trainee selects a theoretical orientation or specialization. They move from learning the basics of therapy to communicating and reasoning like a clinician within a specific framework.

> ★ **Key Takeaway:** Not all AI tools are equally useful in clinical care. Before using one, ask whether the language model has been fine-tuned for your setting, population, or area of practice. Just as a trainee needs specialty training to work effectively, an AI model must be adapted to the right context to be clinically relevant and safe.

How AI Personalizes Care

> ❖ **What You'll Learn:** How Retrieval-Augmented Generation (RAG) allows AI to tailor its responses by drawing on client-specific information, along with the opportunities this creates and the ethical responsibilities it raises.

Most AI models generate responses only from the data they were trained on, which is usually broad and generalized. Clinical care, however, is never generic. To be useful, AI must reflect the individual client's needs, history, and goals.

Retrieval-Augmented Generation (RAG) helps achieve this. RAG pairs a language model with a live search function that retrieves relevant information at the moment of response, such as prior session notes, symptom logs, or treatment plans. This allows the AI to produce outputs that are context-aware, client-specific, and more clinically meaningful.

> **Retrieval-Augmented Generation (RAG)** works by combining a language model with a live search feature. Before generating a response, it can "look up" relevant information from trusted sources. This process improves accuracy and helps reduce the risk of hallucinations.

Think of it like a trainee preparing for a session. They don't rely only on textbook knowledge. They review the case file, recall the client's recent disclosures, and consider treatment goals before deciding how to engage. RAG aims to bring that same level of personalization.

In practice, this could look like:

- A digital assistant summarizing weekly progress by referencing mood logs or past check-ins.
- A chatbot tailoring psychoeducation to reflect a client's presenting issues and stage of treatment.
- A documentation tool recalling a client's own words from earlier sessions to frame a progress note in language that resonates.

These features point toward a future of more responsive and individualized AI support. But with this potential comes clinical responsibility. Any system that pulls from personal data must meet high standards for:

- **Accuracy**, ensuring the retrieved data is relevant and correctly applied.
- **Privacy**, securing sensitive information in accordance with HIPAA or local regulations.
- **Consent**, confirming that clients understand and agree to how their data is used.

The value of RAG systems ultimately depends on how well clinicians guide their use and ensure the outputs align with ethical and therapeutic care.

★ **Key Takeaway:** Retrieval-Augmented Generation helps AI become more responsive by drawing on client-specific data. But personalization must never override privacy, consent, or clinical judgment. Like any tool that handles sensitive information, its value depends on the ethical and thoughtful guidance of the clinician using it.

PART IV: WHAT'S COMING NEXT IN AI

This section looks ahead to the near future, exploring how new developments in AI may reshape both psychological practice and theory. It introduces emerging trends and encourages mental health professionals to start reflecting on their potential impact.

By the end, you will be better able to imagine how AI might grow from stand-alone tools into integrated platforms, recognize new roles psychologists could take on in studying AI behavior, and thoughtfully evaluate technologies such as simulated clients and digital twins from both clinical and ethical perspectives.

How AI Becomes More Holistic

❖ **What You'll Learn:** Today's single-purpose AI tools are expected to develop into integrated platforms that act more like intelligent, domain-specific partners. These systems may enhance how clinicians manage tasks, shift between roles, and keep client care as the central focus.

Today, most AI tools for mental health are still stand-alone. You might use one program to transcribe sessions, another to summarize them, and a separate system to flag risks. On top of that, clients may use journaling apps, chatbots for support between sessions, or billing assistants for insurance. The problem is that these tools rarely talk to each other. They don't connect back to the therapist's goals or clinical reasoning.

In other industries, software has moved from scattered apps to integrated platforms. Mental health is likely to follow the same path. Instead of more isolated tools, the future may bring coordinated, holistic systems that combine many capabilities and are fine-tuned for specific needs, such as trauma recovery, couples therapy, or adolescent care. Rather than functioning as disconnected utilities, these platforms will act more like intelligent collaborators that support care across settings.

Here are four possible examples of what that future might look like (and will be explored in more detail in Chapter 12):

- **AI Co-Pilot Agent:** A live session assistant that transcribes dialogue, tracks emotion in real time, highlights clinical themes, and suggests note content or possible interventions. It acts as an extra set of intelligent eyes and ears to support clinical judgment.

- **AI Workflow Assistant:** A back-office helper that drafts progress notes tied to treatment goals, updates EHR and billing fields, and flags missing documentation. By adapting to a clinician's style, it reduces the paperwork burden while improving consistency.

- **Multi-Modal Therapy Assistant:** A client-facing tool that integrates tone of voice, biometrics like sleep or heart rate, journaling data, and self-reports into a clear picture of mental health between sessions. It can provide real-time coping nudges, alert the therapist to concerning patterns, and feed insights into treatment planning.

- **AI-Driven Relationship Care Platform:** A skill-building system that helps clients work on relational goals outside of therapy. It offers personalized coaching and meaningful engagement between sessions while keeping clinicians updated on progress.

Taken together, these examples suggest a shift from fragmented apps to holistic platforms. The result may not only be greater efficiency but also more continuous, adaptive, and client-centered care.

★ **Key Takeaway:** AI tools are moving beyond single-use features toward integrated systems that support the whole care process. From live session support to reducing paperwork and enhancing between-session engagement, these holistic solutions can help clinicians work more efficiently, personalize care, and stay better connected to their clients' progress.

When AI Becomes the Subject

❖ **What You'll Learn:** Researchers are beginning to look at AI systems not only as tools, but as entities that show their own patterns of behavior. This raises new questions about how machine-generated traits and biases emerge, and how these might shape the way people interact with AI in clinical and everyday contexts.

As AI tools become more advanced and woven into psychological practice, a new shift is taking place. Some researchers are beginning to study the AI itself—not just the tasks it performs, but the way it "behaves."

This area of study, sometimes called *machine psychology*, looks at AI systems as subjects of behavioral research. While AI does not have feelings, consciousness, or intentions, its outputs—such as words, choices, or even errors—can follow patterns that seem strikingly human. For example, language models may respond with such consistency in tone or style that they appear to show recognizable personality traits.

> **Machine Psychology** is the study of AI systems as if they were behavioral agents. It examines how these tools respond in different situations, show consistent patterns, and shape human users' experiences through their tone, style of decision-making, or perceived "personality.

Researchers are now asking questions such as:

- How does an AI system respond when faced with ambiguous or stress-like prompts?
- Can it be influenced to act more cautiously or impulsively depending on the context?
- Do some models show consistent tendencies, such as risk-aversion, confidence, or empathy, that resemble psychological traits?

If training an AI is like teaching a student, then *machine psychology* is the study of how that "student" behaves when tested in different situations.

This is more than theory. The "personality" of an AI can shape trust, engagement, and emotional responses. For example, an overly cautious system might heighten a client's sense of risk, while an overly confident one could encourage misplaced reliance.

Machine psychology provides a framework to understand and manage these dynamics. It helps psychologists anticipate how AI systems might behave and guides ethical, safe, and supportive use in human-centered contexts.

★ **Key Takeaway:** As AI becomes part of mental health care, psychologists are beginning to study these systems as behavioral agents. They show patterns, traits, and biases that influence how people perceive, trust, and interact with them. Machine psychology gives clinicians tools to anticipate these effects and promote more ethical, informed, and supportive use of AI in practice.

Simulated Clients and Digital Twins

❖ **What You'll Learn:** How simulated clients and digital twins are being developed, what they can do, and why they may become important tools for training clinicians and providing more personalized care.

As AI becomes more behaviorally realistic, it is beginning to act not only as a support tool for clinicians but also as a practice patient.

Simulated clients are AI agents designed to mirror specific mental health presentations. They are increasingly being used in training to help students and early-career clinicians build skills such as:

- Conducting clinical interviews
- Assessing risk or suicidality
- Clarifying diagnostic impressions
- Practicing crisis de-escalation

> **Simulated Client:** AI-powered "practice patients" that mirror real mental health presentations. They give students and trainees a safe, repeatable way to build key clinical skills, including risk assessment, diagnostic clarification and crisis response.

These AI agents act like digital standardized patients. Instead of being played by actors, they draw on real-world clinical data. They can adjust affect, mimic trauma symptoms, and respond dynamically to questions, creating rich and structured training opportunities.

A related but more personalized tool is the *digital twin*. While simulated clients are designed for general training, a digital twin is built from an individual client's data, such as clinical notes, physiological measures, and engagement history. When combined with Retrieval-Augmented Generation (RAG), digital twins can draw on real-time context to provide highly tailored responses.

> **Digital Twin:** An AI model built from an individual client's real data. It can reflect clinical patterns, track changes over time, and support personalized care strategies between sessions. However, it also raises important questions about privacy, consent, and trust.

In practice, digital twins are being explored to:

- Simulate how an individual client might respond to different treatments
- Detect and track subtle psychological or behavioral changes
- Provide personalized support between sessions, based on prior patterns

These possibilities bring clear promise but also new ethical challenges. Who owns the data? Has the client given meaningful consent? Could such systems change the therapeutic alliance or how care is experienced? These issues will be examined further in the chapter on regulation, privacy, bias, and ethics.

★ **Key Takeaway:** AI-powered simulated clients and digital twins open new possibilities for training and personalized care. At the same time, they raise important questions about identity, consent, and data ethics. Mental health professionals will be essential in guiding how these tools are used and ensuring that clients remain protected.

Conclusion

Artificial intelligence is beginning to change how mental health professionals work, but not what it means to be one. Tools that analyze language, track behavior, or assist with documentation are expanding what's possible in care. Yet AI cannot replace the heart of the profession: empathy, ethics, cultural awareness, and lived experience.

In this chapter, we explored how AI learns, generates responses, and adapts to different contexts. We looked at both its strengths and its limits. Throughout, we returned to one guiding metaphor: AI is like a student. It may be fast-learning, capable, and helpful, but it still needs supervision. Like any trainee, AI requires guidance, oversight, and a clearly defined role.

It's not enough to know what AI can do. We must also understand how it works, where it falls short, and what it means for practice. These systems can hallucinate, misinterpret tone, or use clinical terms out of context. In high-stakes settings, such mistakes can affect outcomes, trust, and client safety.

Mental health professionals do not need to become engineers. But developing fluency with AI is part of ethical, evidence-based practice. That fluency means knowing what a tool was trained on, whether it fits your population, and if its outputs are fair, accurate, and clinically useful. Ask questions such as:

- What kind of data was this tool trained on?
- Is it appropriate for my clients?
- Does it support transparency, fairness, and clinical relevance?

When used wisely, AI can reduce burden, improve access, and offer meaningful support. But it must serve clinical care, not lead it. Mental health professionals remain the ethical filter, the decision-maker, and the human presence at the center of healing. AI can assist, but the work, ultimately, remains human.

If AI Were a Psychology Student: A Working Metaphor

To explain the fundamentals of AI, we have used the metaphor of a developing psychology student—learning from examples, practicing skills through repetition, and always needing guidance. The table below summarizes this comparison. It highlights how AI, like a trainee, can be useful but also limited. This lens can help mental health professionals evaluate AI systems with the same care and supervision they would give to a student in training.

Student Trait	AI Equivalent
Learns from supervision	Trained on labeled data (e.g., supervised learning models that improve with expert-reviewed input)
Gains experience over time	Improves with exposure to more—and better—examples (e.g., analogous to machine learning models refining predictions with larger and more diverse datasets)
May make premature conclusions	Generates outputs based on statistical patterns, but may lack understanding of context, leading to errors when encountering novel or underrepresented data
Can misread nuance or context	Lacks emotional understanding and lived experience, meaning it may misinterpret tone, sarcasm, or cultural nuances
Benefits from diverse client exposure	Needs representative, high-quality training data, including diverse demographic and psychological contexts, to avoid bias and overgeneralization
Risks overconfidence without oversight	May produce plausible sounding but incorrect outputs (e.g., AI Co-pilots flagging shifts in patient behavior during therapy)
Must explain clinical reasoning	Requires explainability (XAI) to justify predictions (e.g., AI-powered decision support in clinical psychology)
Needs supervision and ethical boundaries	Depends on human-in-the-loop systems for safe deployment (e.g., AI workflow assistants managing clinician tasks)

As we look ahead, the next chapter explores the other half of the AI equation: data. If AI is the engine, data is the fuel, and what goes into the system will shape everything that comes out. Understanding the types, quality, and ethical handling of data is essential to ensuring that AI enhances, rather than undermines, psychological care.

Chapter 3: The Role of Data

Artificial intelligence may sound advanced, but at its heart it is like a student, and data is the lesson plan. AI does not think or reason as people do. Instead, it looks for patterns in past examples and then tries to apply those patterns to new situations. This means the quality, variety, and ethical standards of the training data shape everything the system produces.

In mental health, this has direct consequences. The data used to train an AI model can influence whether a diagnosis is accurate, whether risks are identified correctly, and whether care is fair and inclusive. A model trained on biased or narrow data may look effective at first, but it can fail when applied to your client.

This chapter explores the kinds of data behind AI systems. We will look at structured sources such as diagnostic codes, and unstructured sources such as speech, biometrics, and digital behavior. We will also consider how poor data can distort results, how bias enters the process, and what clinicians can do to spot risks and support safer, more inclusive tools.

The purpose is not to make you a data scientist. The aim is to give you a practical way to judge whether an AI tool rests on solid evidence or shaky assumptions. If you do not know what went into the model, you cannot fully trust what comes out.

What Types of Data Power AI in Mental Health

❖ **What You'll Learn:** You'll learn about the main types of data that train AI in mental health, such as clinical, behavioral, cognitive, and experimental data. You'll also see why knowing the differences between these types of data is important when evaluating AI tools.

AI tools in mental health draw on many kinds of information. This includes not only test scores or diagnoses, but also language, behavior, and physical signals. These inputs usually fall into two groups:

- **Structured data** is organized, labeled, and easy for machines to read, such as intake forms or diagnostic codes.

- **Unstructured data** is more complex and filled with context, such as therapy notes or spoken language.

Both are important. Structured data often shows *what* happened, while unstructured data can give insight into *why* it happened. The strongest AI tools combine both to create a fuller picture of the person and their situation.

Next, we will look at the main types of data AI systems use in mental health.

Clinical and Diagnostic Data

We start with clinical and diagnostic data. Examples include electronic health records (EHRs), standardized tools like the PHQ-9 or GAD-7, intake forms, and therapy notes. These sources are often structured and labeled, which makes them easier for AI systems to process.

In this context, the labels—sometimes called ground-truth labels—are the correct answers or outcomes identified by people or verified systems. They are used to train AI models and to check whether the model's predictions are accurate. Without ground-truth labels, it is hard to know if an AI system is actually learning or producing reliable results.

> **A ground-truth label** is the correct answer given to train an AI model. For example, it might involve tagging a note as indicating depression. This label shows the system what the "right" outcome looks like.

AI models that use clinical data can help in several ways:

- Spotting patterns in how symptoms change over time
- Predicting how a person might respond to treatment
- Flagging risks based on diagnostic history

The value of this data depends on how complete and consistent it is. Missing information or differences in how clinicians write notes can create problems. With enough high-quality and diverse training data, well-designed models may be able to filter out some of these inconsistencies.

Behavioral and Biometric Data

With the growth of mobile apps and wearable devices, AI tools now collect real-time data on sleep, movement, heart rate, and other behavioral signals. They can also track digital patterns such as screen time, typing speed, or app use, which may reflect changes in mental health.

Typical sources include:

- **Wearables and smartphones**, which monitor sleep, movement, and screen use
- **In-app activity**, such as logins, messaging, or how long someone stays in a session

For instance, poor sleep and less physical activity may be linked to depressive episodes. Still, this information must always be considered in a clinical context, since data on its own does not equal a diagnosis.

Cognitive and Affective Data

AI can study how people speak, write, or show emotion to draw conclusions about mood, attention, or possible cognitive decline. These methods are increasingly being used to support assessment.

Important sources include:

- Speech patterns, such as tone, pacing, and pauses
- Facial expressions, captured on video or with sensors
- Eye movements or gestures, sometimes tracked in labs or virtual reality settings

These signals can highlight distress, disengagement, or inconsistency during evaluations. However, they must always be interpreted with awareness of cultural and personal differences, or there is a risk of misclassification.

AI can also use sentiment analysis (introduced in Chapter 2) to study written or spoken language, identifying emotional tone or tracking changes over time. For example, it might detect hopelessness in a journaling app. Still, these systems rely on training data that reflects the right cultural context. Without it, they may generate false positives or overlook important warning signs.

Research and Experimental Data

Surveys, experiments, long-term studies, and observational research create valuable datasets that can be used to train or test AI models, especially when the data is labeled or standardized.

These datasets can be applied to:

- Predicting population-level mental health trends
- Simulating possible treatment responses
- Comparing psychological traits or symptoms across cultures

While research data is often high quality and carefully collected, it may not always represent the full diversity of real-world populations. Because of this, AI models trained on such data may not perform well across different cultural, demographic, or clinical settings.

Taken together, data from clinical, behavioral, biometric, emotional, and experimental sources provide a layered perspective on mental health. At the same time, they add complexity. Each type of data has its own strengths, limitations, and privacy concerns. For mental health professionals, understanding these differences is key when evaluating AI-powered tools.

Next, we will look at an important distinction in AI training: the difference between structured and unstructured data.

★ **Key Takeaway:** Knowing the kinds of data that AI systems use allows mental health professionals to better judge whether these tools are accurate, relevant, and appropriate for client care. Without this understanding, it is easy to miss hidden biases or place too much trust in what AI can deliver.

Structured vs. Unstructured Data

❖ **What You'll Learn:** You'll learn the difference between structured and unstructured data, and why both are important when assessing how mental health AI tools work and where they may have limitations.

As discussed earlier, structured data tells you *what* happened; unstructured data may help explain *why*. Both are essential in psychological AI.

Understanding the difference helps clarify how AI systems are trained, where they work well, and where they may go wrong. Each type has specific strengths, limitations, and ethical considerations that affect clinical reliability.

Structured Data: Organized, Labeled, and Machine-Friendly

Structured data is organized into predefined fields such as checkboxes, dropdown menus, or numerical scales. This makes it easier for AI systems to analyze and compare information across clients or over time.

Structured Data fits into predefined formats such as checkboxes, dropdown menus, or numerical scales. It is straightforward for AI to process and compare, but it often lacks depth and does not capture emotional nuance.

Examples of structured data include:

- Diagnoses coded with ICD-10 or DSM-5-TR
- Standardized assessments such as the PHQ-9 or MMPI-2
- Electronic health record fields like demographics, medications, or treatment history

Because structured data is clear and consistent, it is often used to train models that track symptom changes, predict treatment outcomes, or identify clinical risks. Its uniformity makes it scalable and efficient.

The limitation is that structure can reduce depth. These data points may show what is happening but miss how clients describe their experiences. For example, a low PHQ-9 score signals distress, but it does not capture what that distress feels like or how it shapes daily life. AI models trained only on structured inputs may lose important context or oversimplify complexity.

This is where unstructured data plays an important role.

Unstructured Data: Complex, Rich, and Harder to Interpret

Unstructured data is less organized but far more expressive. It includes open-ended content such as written notes, spoken language, video, or social media. This type of data captures how people think, feel, and communicate in daily life, often providing nuance, tone, and context that structured formats leave out.

Unstructured Data includes open-ended sources such as text, speech, or video. It captures tone, context, and emotion, but it is harder for AI to interpret and is especially sensitive to cultural differences.

Examples of unstructured data include:

- Therapist session notes and client journals
- Audio recordings from therapy sessions or interviews
- Language patterns on social media
- Recordings of facial expressions or gestures

To interpret this kind of data, AI systems use methods such as natural language processing (NLP), computer vision, and sentiment analysis. These tools can identify emotions, shifts in narrative, or nonverbal cues that structured data would miss.

Interpretation, however, is rarely simple. Tone, expressions, and behaviors differ across cultures, languages, and personal styles. Without diverse training data, AI systems may misinterpret sarcasm, fail to notice distress, or flag harmless language as a risk.

Unstructured data provides depth but also brings complexity. When used responsibly, it can help AI capture the richness of human experience. This requires cultural sensitivity, context awareness, and strong ethical safeguards.

Recognizing the strengths and limits of both structured and unstructured data is essential. But what happens when the data itself is flawed, incomplete, or biased? That is where the issue of data quality becomes central.

★ **Key Takeaway:** Structured data is simple for AI to analyze but often lacks emotional depth. Unstructured data captures richer human signals but is harder to interpret and more vulnerable to cultural misreadings. Understanding both is essential when judging whether an AI tool can address the complexity of real-world mental health.

Data Quality

❖ **What You'll Learn:** You'll learn why data quality is a key factor in helping to determine whether AI tools are accurate, ethical, and clinically useful. You'll also see how poor data can cause real-world harm in mental health care.

As discussed in Chapter 2, AI systems are not neutral. They take on both the strengths and weaknesses of the data they are trained with. In this section, we will look more closely at what makes data high or low quality, and why that difference is so important for clinical safety.

Poor-quality data does more than reduce accuracy. It can create biased predictions, unreliable risk assessments, and ethically troubling outcomes. For clinicians, knowing what counts as "good data" is an essential step in judging whether an AI tool can be trusted in practice.

Why Bad Data Leads to Bad Predictions

AI models learn by detecting patterns in past data. If the training data is incomplete, mislabeled, or inconsistent, the model will carry those flaws forward—sometimes with confidence. This can lead to:

- Underestimating risk in marginalized populations
- Misclassifying symptoms that appear in less typical ways
- Reinforcing outdated or biased clinical assumptions

For example, if an AI system is trained on records from a clinic that has historically underdiagnosed trauma in men, it may learn to overlook signs of trauma in male clients. The system is not biased on purpose; it simply inherits the blind spots of its training data, much like a student taught with flawed examples.

Common data quality problems in psychology include:

- **Incompleteness:** Missing values, gaps in session notes, or lack of follow-up data can distort the clinical picture.
- **Noise:** Irrelevant or inconsistent details, such as varied note styles or transcription errors, can obscure meaningful patterns.
- **Mislabeling:** Incorrect diagnoses or outdated terms can teach the model faulty associations that continue to affect its predictions.
- **Lack of standardization:** Differences in how clinicians document symptoms or outcomes make it difficult for AI to recognize consistent patterns across settings.

These are not just technical issues. They can directly affect care. A mislabeled or misinterpreted piece of data can result in decisions that misguide treatment for a client.

Overfitting and Underfitting: When AI Misses the Mark

Even with high-quality data, AI models can still fail if the learning process is flawed. Two common problems are overfitting and underfitting:

- **Overfitting** occurs when a model memorizes the training data instead of learning general patterns. It may look accurate during testing but break down when faced with real-world variety. In mental health, this could mean a model trained on one cultural group fails to recognize valid expressions in another.

- **Underfitting** happens when a model is too simple or trained on data that lacks diversity. It never learns the important patterns. For example, an emotion-recognition tool trained only on exaggerated examples may perform poorly with real faces in different conditions.

These problems are not just technical. They shape clinical usefulness. A model that overfits might over-predict distress in some groups, while one that underfits may fail to detect it at all.

Practical Example: Early PTSD Detection

Consider an AI model designed to detect early signs of post-traumatic stress disorder (PTSD) from therapy transcripts. If its training data comes mainly from military veterans, the model may learn to link PTSD primarily with combat-related language and symptoms. When applied to survivors of domestic violence or refugees, it could then miss equally valid expressions of trauma.

The problem is not the model itself but the data it was trained on. An AI system can only reflect the patterns it has seen. Without a dataset that is diverse and representative, even a well-built model may produce biased or misleading results.

What This Means for Mental Health Professionals

When using an AI tool or reviewing its results, it is important to ask:

- What kind of data was used to train this model?
- Does the data represent the population I serve?
- Were there quality checks to ensure accuracy?

High-quality data allows AI to strengthen clinical judgment. Poor-quality data, no matter how sophisticated the algorithm, can mislead, create risks, or weaken trust in care.

Next, we will look at a key part of data quality—representation—and why inclusive training datasets are essential for building AI tools that serve all populations fairly.

★ **Key Takeaway:** An AI tool is only as reliable as the data it was trained on. Incomplete, mislabeled, or biased data can lead to flawed predictions that influence care decisions. Mental health professionals should ask careful questions about a tool's training data before relying on its results.

Bias in AI

❖ **What You'll Learn:** You'll learn how bias makes its way into AI systems used in mental health care, how it shows up in real-world practice, and what clinicians can do to reduce harm through careful evaluation and ethical use.

Bias in AI is not always obvious, but its impact can be significant. In mental health care, biased systems may reinforce diagnostic blind spots, misclassify symptoms, or deepen inequities in treatment. Because AI learns from historical data, it can easily repeat past errors unless the training sources are diverse, transparent, and carefully reviewed. To use AI tools responsibly, mental health professionals need to understand how bias develops and how it appears in real-world practice.

Where Bias Begins

Most AI models are not biased on their own, but they learn bias from the data provided to them. If historical records contain systemic disparities or if certain populations are underrepresented, those patterns can shape how the model functions.

Common sources of bias include:

- **Historical inequalities** embedded in clinical records or legal data
- **Sampling bias**, where one group is overrepresented in the training data
- **Labeling bias**, where subjective clinical judgments are treated as "ground truth"

These challenges are not only technical. They also reflect larger social and institutional factors that shape how data is collected, interpreted, and used.

Examples of AI Bias in Mental Health Contexts

Bias in AI may sound abstract, but in mental health and behavioral settings it can have very real, and sometimes high-stakes, effects. Below are examples showing how well-performing models can fail when training data lacks diversity, cultural context, or developmental nuance.

Suicide Risk Prediction Using Wearables

AI tools that monitor physiological data—such as heart rate, sleep, and activity logs—are being tested to flag suicide risk. A 2025 pilot study (Um et al.) reported strong predictive accuracy with commercially available devices. However, the dataset was small and homogenous, raising concerns about how well the model would work for broader populations. These tools also miss important context, such as trauma history or social factors, that cannot be seen in biometric signals alone.

Facial Emotion Recognition in Autism Diagnosis

Most emotion-recognition AI is trained on neurotypical faces. A 2024 review (Banos et al.) found that such models often misinterpret or overlook expressions common among autistic individuals. Without sensitivity to tone, gesture, or context, these tools risk misclassifying behavior and undermining diagnostic accuracy for neurodivergent clients.

AI-Based Hiring in Organizational Psychology

Amazon's widely cited hiring algorithm penalized résumés from women because it had been trained on male-dominated historical data. Words like "women's college" were down-ranked (Chang, 2023). While not a clinical case, it shows how AI can reinforce inequities—an important lesson for I/O psychologists considering similar tools in workplace assessments.

Depression Detection in Social Media

AI models trained on Western, English-language posts have struggled to identify depression in users from the Global South. A 2024 study (Ali et al.) found misclassification tied to cultural differences in language and self-expression. This raises concerns about whether so-called "universal" tools are effective in multilingual or cross-cultural mental health contexts.

Dialect Bias in Language Models

Speech recognition systems often perform poorly when transcribing African American Vernacular English (AAVE). A 2024 study (Zolnoori et al.) showed that errors sometimes affected clinically meaningful phrases, potentially distorting therapy transcripts or screenings. The problem stems from underrepresentation in training datasets.

These examples highlight why clinicians must ask what an AI system was trained on, and who may have been left out. When data does not reflect real-world diversity, even advanced tools can produce poor or misleading outcomes.

Approaches to Mitigate Bias in AI

Bias in AI is not simply a glitch. It often comes from choices made during design and from limits in the data. As discussed in Chapter 2, these biases can lead to serious consequences in care, including misdiagnosis and inequitable treatment. Bias cannot be removed completely, but it can be reduced, managed, and made visible with the right strategies.

In mental health care, where nuance is critical and the risks of false positives or missed signals are high, addressing bias is not optional. It is both a clinical and ethical responsibility.

Four important techniques for reducing bias in mental health AI tools include:

- **Bias Audits:** Regularly checking model outputs across demographic groups such as race, gender, age, neurotype, and language to spot performance gaps.

- **Fairness Testing:** Ensuring that people with similar profiles receive consistent results, regardless of background. This can reveal subtle inequities.

- **Inclusive Data Practices:** Actively including underrepresented groups and clinical presentations in training data, with variation in language, geography, symptom expression, and cultural norms.

- **Human Review Loops:** Building systems where clinicians remain in control, able to override, contextualize, or question AI outputs based on client history, lived experience, or therapeutic insight.

Clinicians do not need to become machine learning experts, but they should serve as informed advocates. This means asking:

- Was the model tested on the populations I work with?
- Does the tool share its known limitations?
- Can I step in when I believe the AI is wrong?

Bias mitigation is not just a task for developers. It is a shared duty. Mental health professionals have the authority to push for transparency, call for ethical design, and keep client wellbeing at the center.

The next section explores one of the most sensitive challenges in AI-driven mental health care: protecting privacy, honoring consent, and upholding ethics when personal data becomes fuel for prediction. As AI systems absorb more personal and behavioral data, the focus shifts from "Is this accurate?" to "Is this safe, fair, and just to use?"

★ **Key Takeaway:** AI bias may not be obvious, but it can directly impact the quality of care, especially for marginalized groups. Mental health professionals play an essential role in spotting biased outputs, asking critical questions, and advocating for tools that uphold fairness and cultural relevance in clinical practice.

Data Privacy and Consent

❖ **What You'll Learn:** You'll learn why protecting mental health data requires more than basic privacy measures, and how clinicians can assess whether AI tools meet both ethical and legal standards.

Psychological data carries deep ethical importance. It is more than health information—it represents identity, lived experience, and emotional history. As AI becomes more integrated into mental health care and assessment, professionals must ensure that client data is gathered, used, and stored in ways that protect privacy, autonomy, and trust.

AI can help identify risks, make workflows more efficient, and tailor care to individual needs. But if it undermines trust, transparency, or confidentiality, its value is quickly lost.

Why Mental Health Data Is Uniquely Sensitive

Not all data is the same. Mental health data reflects some of the most personal aspects of a person's life—memories, emotions, relationships, and vulnerabilities. Unlike blood pressure or cholesterol levels, this information is highly contextual and emotionally charged.

AI tools in this field may analyze:

- Therapy transcripts or session notes
- Psychological assessments and diagnoses
- Voice tone, facial expressions, or gestures
- Passive data from wearables or mental health apps
- Journals, chat logs, or mood-tracking entries

Once digitized, this information can be stored, copied, analyzed, or even sold. If not properly safeguarded, it can expose clients to risks that extend far beyond a privacy breach, including misdiagnosis, stigma, or legal and employment consequences.

For this reason, mental health data requires stronger protections. Standard HIPAA or GDPR compliance may not be enough when AI relies on this level of personal detail. Clinicians must think not only about how the data is collected and stored, but also how it is interpreted, who can access it, and whether clients truly understand where their data may go.

Regulatory Blind Spots

Laws such as HIPAA in the U.S. and GDPR in Europe establish important standards for protecting health data, focusing on consent, transparency, and limiting how much data is collected. Yet these frameworks were written in an era before the unique risks and requirements of AI-driven mental health tools were imagined.

Many mental health apps, digital platforms, and third-party services operate outside traditional clinical settings. This often places them in legal gray areas where HIPAA or GDPR may not apply—or may be weakly enforced. These tools can collect mood logs, speech samples, sleep data, or journaling entries without the oversight required in a licensed therapy office.

This gap is not just about technology—it directly affects clinical decision-making.

Mental health professionals now encounter tools that are:

- Lightly regulated despite handling sensitive client information
- Unclear about how data is stored, shared, or reused
- Difficult to evaluate for compliance or ethical protections

Recommending or using an AI-powered app is more than a technology choice—it is a clinical judgment. If a tool collects client data, clinicians should ask:

- Is the data encrypted and securely stored?
- Do clients give meaningful consent for how their data is used or shared?
- What happens to the data after collection—does it train future models, get sold, or remain private?

Until regulation adapts, it falls to practitioners to close this gap—choosing tools carefully, explaining risks openly, and protecting client trust beyond the minimum legal requirements.

Clinician Responsibility: A Call to Action

Mental health professionals are not only caregivers—they are also gatekeepers of the digital tools entering therapeutic practice. As AI-powered apps, platforms, and assessments become more common, clinicians are on the front lines of deciding which tools are safe, ethical, and appropriate for client care.

This responsibility goes beyond understanding what an AI tool does. It requires asking what it demands from clients.

Before introducing a digital mental health tool, clinicians should consider:

- **Is the tool transparent?** Does it explain clearly what data it collects, how that data is stored, and what happens to it afterward?

- **Can clients truly opt out?** If participation is required to access care, then consent is not fully voluntary.

- **Is there accountability if harm occurs?** If the tool mislabels, misfires, or mishandles sensitive data, what protections exist for clients?

AI tools may be new, but the clinician's duty remains the same: to protect client wellbeing, autonomy, and trust. That means asking tough questions, carefully examining claims, and advocating for ethical standards—even when a tool is marketed as cutting-edge or arrives with pre-approval.

In a rapidly changing digital landscape, it is clinical judgment, not the software, that must lead the way

★ **Key Takeaway:** Mental health professionals cannot hand off responsibility for privacy and consent to developers or platforms. They must actively assess whether an AI tool safeguards client data, respects informed consent, and upholds ethical and legal standards—even when the technology sits outside traditional clinical systems.

Conclusion

AI systems are only as reliable as the data that shapes them. In mental health, where precision, empathy, and ethics are essential, the data behind an AI tool is not just technical—it is clinical. It shows who was included, who was left out, and what assumptions influence the system.

When designed well, AI can help identify risks earlier, personalize care, and increase access to services. But when trained on biased, narrow, or poorly understood data, the same tools can misdiagnose, miss critical warning signs, or widen existing disparities.

This is why clinicians cannot be passive users. They are decision-makers who determine whether a tool should be part of practice. Knowing the difference between structured and unstructured data, spotting signs of poor data quality, and asking difficult questions about transparency, bias, and consent are now core responsibilities of ethical care.

As AI takes on a larger role, the responsibility of mental health professionals grows alongside it. The future of psychological AI will not be shaped by algorithms alone, but by the values and choices of the people who decide how it is used.

Supervision Analogy: Training AI Like a Psychology Student

As a final recap on data, think of it this way: if AI is still learning, we are the ones setting the syllabus.

If AI is the student, data is its curriculum, and clinicians serve as its supervisors. Good supervision means paying attention not only to outcomes, but also to the materials, methods, and values that shape the learning process.

If you would not...	Then don't with AI...
Let a student diagnose without supervision	Use AI predictions without human review
Teach only one cultural presentation	Train models on homogenous data
Skip consent before recording sessions	Use client data without clear, informed consent
Expect a trainee to perform across specialties	Assume one model works across all populations
Rely on flawed case notes to teach judgment	Feed models low-quality, mislabeled, or incomplete data
Ignore cultural factors in supervision	Dismiss cultural context in data collection and interpretation
Assume learning is automatic without reflection	Trust AI learning without bias audits or quality controls
Let a student cite made-up studies	Accept hallucinated citations or AI-generated content uncritically

In the chapters ahead, we will look at how these issues play out in specific applications, from clinical care to forensic evaluations. In every case, the same principle applies: the power of AI begins—and is constrained—by the quality and integrity of its data.

Chapter 4: AI in Clinical and Counseling Practice

Mental health professionals in both clinical and counseling roles are facing increasing demands. Waitlists are growing longer, caseloads are expanding, and documentation tasks are piling up. At the same time, client needs are becoming more complex. Whether helping someone through a life transition or diagnosing overlapping conditions, clinicians must handle a wide range of challenges, often with limited time and resources.

This chapter looks at how artificial intelligence (AI) is being used to support both counseling and clinical work. Clinicians are working to engage clients from diverse backgrounds, keep momentum between sessions, reduce paperwork, and make care decisions that are both timely and personalized.

Many AI tools serve more than one purpose, so you may see some of them appear in multiple sections. For instance, a chatbot that boosts client engagement might also expand access to care or assist with early risk detection.

In addition, in this and the following chapters (4–9), we reference specific tools, platforms, and companies to illustrate how AI is being applied in different areas of psychological practice. These examples were selected for relevance, visibility, or presence in peer-reviewed research, and are included for informational purposes only—not as endorsements or claims of effectiveness. At the time of publication, the authors have no financial interest in any of the tools mentioned, except for Relationship Workout, LLC.

Finally, readers interested in innovations in cognitive and neuropsychological assessments should also see the AI in Neuropsychology section of Chapter 8: AI in Health Psychology.

Addressing Limited Access to Mental Health Services

Access to mental health care continues to be one of the biggest challenges in the field. Many people face long waitlists, a shortage of clinicians, high costs, and limited availability in their area. These problems are especially common in rural and underserved communities (Babu & Joseph, 2024; WHO, 2021). On top of that, stigma, lack of culturally appropriate services, and language barriers make it even harder for marginalized groups to get the care they need.

Opportunities and Benefits

AI tools are starting to help break down some of the barriers to accessing care. Culturally aware chatbots and multilingual self-help apps are examples of low-cost, private, and scalable options that can help people begin getting support— especially those who might never connect with a clinician through traditional means.

When these tools are built with equity, privacy, and clinical quality in mind, they can close access gaps without adding more work for providers. They're available 24/7, adjust to users' language and cultural needs, and can screen for issues early or monitor passively to encourage timely help-seeking. Here are a few ways these tools are expanding access:

24/7 Support Without Location Limits
An AI chatbot like Wysa is available at all hours on smartphones. This makes it easier for people to get support no matter where they are, which is especially helpful in rural or remote areas where providers are hard to reach.

Helping More People Without Adding to Provider Workload
These tools can support many users at once by offering basic education, coping tools, and symptom tracking. In low-resource settings, they can reduce waitlists and free up clinicians to focus on clients with more complex needs.

Personalized Support for Different Languages and Cultures
AI platforms like Tess and Wysa use language processing to adapt the tone and content of their responses based on the user's background. This helps make the experience more relevant and builds trust, particularly in communities that are often underserved.

Support When Human Help Isn't Available

When providers are unavailable, such as after hours or during staff shortages, AI tools can offer immediate emotional support. They can walk users through grounding exercises, provide cognitive reframes, or check in during difficult moments.

A Lower-Stakes Way to Seek Help for the First Time

For people who feel unsure or uncomfortable about therapy, AI tools can offer a safe, private starting point. This can be an important first step for those who might otherwise avoid seeking help altogether.

Example Applied AI Tools

A variety of AI tools are already helping to lower barriers to mental health care, especially for people who have trouble accessing traditional services. Tools like multilingual chatbots and anonymous, around-the-clock platforms show how AI can make care more accessible by being affordable, available at any time, and sensitive to cultural needs.

Wysa App
- **What it does**: A mental health app combining CBT, mindfulness, and dialectic behavior therapy (DBT) tools with an AI-powered chat interface.
- **In this context**: Designed for anonymous, stigma-free use, Wysa allows users to explore challenges on their own terms—without needing formal diagnosis or referral.
- **Why it matters for access**: Removes stigma and enables first-time help-seeking, especially in populations less likely to pursue traditional therapy.
- **Evidence**: Clinical evaluations found Wysa reduced symptoms of depression and anxiety and was rated highly for accessibility and cultural relevance (Babu & Joseph, 2024).

Tess
- **What it does**: Tess is an AI-powered emotional support chatbot used by health systems, insurers, and universities.
- **In this context**: Offers multilingual support across short message service (SMS), apps, and messaging platforms—reaching people where they are, in the language they speak.
- **Why it matters for access**: Extends the reach of care to underrepresented language communities and helps fill staffing gaps at scale.

- **Evidence**: Studies show Tess improved engagement and emotional well-being among high-need and underserved populations (Abd-alrazaq et al., 2020).

Relationship Fitness for Men AI Relationship Coach
- **What it does**: A chatbot-based coaching app tailored for men, offering anonymous, voice-enabled emotional support focused on relationships.
- **In this context**: Delivers on-demand support without requiring formal counseling entry, addressing stigma that prevents some men from seeking help.
- **Why it matters for access**: Offers an entry point for a group historically underrepresented in therapy settings.
- **Evidence**: Research shows that AI-based conversational agents can serve as low-barrier entry points for populations deterred by stigma or structural obstacles. Men who are hesitant to seek traditional therapy have reported increased emotional reflection, perceived support, and readiness to engage with counseling after using such tools (Li et al., 2023; Siddals et al., 2024).

Practical Example: AI Chatbots Expanding Access to Underserved Populations

AI-powered chatbots and smartphone apps are increasingly seen as promising tools for closing gaps in mental health care, especially for underserved and vulnerable populations such as those in rural areas, lower-income communities, or with limited access to clinicians. For example, Pozzi and De Proost (2024) emphasize that CBT-based chatbots and apps are being used precisely because of the shortage of mental health professionals, and because these chatbots can offer support at times when human help is unavailable. Such tools offer relatively low cost, anonymity, and flexibility in timing that help reduce traditional barriers like cost, transportation, and stigma (Haque & Rubya, 2023).

Empirical research also indicates that chatbot interventions can yield positive outcomes in populations that face access constraints. For instance, a recent two-phase observational study found that a ChatGPT-powered chatbot using CBT techniques reduced anxiety symptoms by more than 20% across both phases, with participants reporting high satisfaction with its accessibility and personalization (Manole, Cârciumaru, Brînzaş, & Manole, 2024).

However, authors caution that chatbots are not a full substitute for human care. As Balcombe (2023) notes, while chatbots can improve access and engagement, concerns remain about quality control, cultural sensitivity, and the ability to safely handle crisis or high-risk situations—highlighting the need for human oversight and ethical safeguards.

Concluding Remarks

AI can help extend the reach of counseling services, especially for those struggling to stay engaged or access care in traditional ways. When designed thoughtfully, these tools offer clients low-stigma, always-available pathways to reinforce progress, build insight, and stay connected to therapy. In this context, AI is not about automation, but amplification: helping people do more of the emotional work they already want to do, with support that fits their lives.

Therapeutic Engagement & Continuity

Clients often struggle to maintain motivation and complete therapeutic tasks between sessions. High dropout rates, skipped homework, and inconsistent follow-through hinder progress and reduce the long-term effectiveness of treatment, especially when ongoing engagement tools are lacking, or clinicians have limited capacity to provide support outside sessions.

Opportunities and Benefits

AI tools are emerging as practical supports for sustaining therapeutic momentum between sessions. By offering personalized nudges, on-demand skill practice, and real-time behavioral feedback, these technologies help clients stay connected to their goals outside the therapy room. Rather than replacing clinicians, AI systems extend the reach of care—reinforcing routines, flagging early signs of disengagement, and making follow-through more achievable in daily life.

Reinforcing Skill Use Between Sessions

AI-powered chatbot like Wysa guides users through evidence-based practices such as journaling, cognitive reframing, or mindfulness, helping clients apply therapeutic tools in real time when issues arise.

Delivering Personalized Nudges

Systems can tailor reminders or suggestions based on mood, engagement patterns, or progress. These nudges encourage clients to complete homework, reflect on their progress, or stick to routines without therapist prompting.

Boosting Motivation Through Gamification

Progress tracking, badges, and micro-rewards can make therapeutic tasks feel more tangible and rewarding, particularly for younger clients or those with low initial motivation.

Identifying Disengagement Early

AI tools can detect signs of client withdrawal—such as skipped tasks, mood stagnation, or reduced app usage—and alert therapists to intervene before dropout occurs.

Making Follow-Through Easier and More Frequent

By lowering the friction of accessing support tools, AI helps integrate therapy into daily routines. Quick exercises, structured prompts, or in-the-moment check-ins support consistency even during busy or stressful periods.

Example Applied AI Tools

A growing number of AI tools are being used to enhance therapy engagement and help clients stay connected to the therapeutic process. These technologies support motivation, consistency, and personalization—often outside traditional face-to-face sessions.

Note: Some AI tools discussed here also appear in the previous section on access to care. This reflects their dual role—not only in lowering entry barriers but also in helping clients stay connected and engaged throughout treatment. Their versatility makes them especially valuable in expanding and sustaining mental health support.

Wysa

- **What it does:** Combines AI-guided conversations with CBT, mindfulness, and DBT tools to help users process emotions and build coping skills.
- **In this context:** Offers anonymous, on-demand support that encourages daily engagement and reinforces therapy goals outside session time.
- **Evidence:** Clinical evaluations report reductions in stress and anxiety, with high engagement and usability ratings across diverse populations (Babu & Joseph, 2024).

Therabot

- **What it does**: A large language model–based mental health chatbot that delivers structured CBT exercises and personalized interventions for depression, anxiety, and eating disorders.
- **In this context**: Offers 24/7 access to therapeutic guidance and helps clients maintain progress between appointments through structured self-guided work.
- **Evidence**: In a randomized controlled trial, Therabot users reported significantly greater reductions in depression and anxiety symptoms than those on a waitlist control. Engagement was high, with 96% completing at least one session per week (Thompson et al., 2024).

Relationship Fitness for Men

- **What it does**: A mobile app and online platform featuring a voice-enabled AI "relationship coach" chatbot and aligned AI-enabled tools, designed to help men assess, track, and strengthen their relationship skills.
- **In this context**: Prompts users to regularly rate their relationship and journal their experiences. These entries—when shared with a therapist—offer valuable insight into daily experiences and patterns between sessions.
- **Evidence**: Research indicates that AI chatbots can help users build consistent self-reflection habits and sustain therapeutic engagement between sessions. In particular, studies have found that tools designed for populations traditionally less likely to seek traditional therapy—such as men—can improve follow-through on emotional tasks and increase ongoing commitment to care (Li et al., 2023; Siddals et al., 2024).

Therapist Dashboards with Engagement Analytics

- **What they do:** Track user behavior across therapeutic tasks (e.g., mood check-ins, journaling, exercise completion) and flag early signs of disengagement.
- **In this context:** Help clinicians proactively respond when clients begin skipping tasks or withdrawing, improving retention and treatment consistency.
- **Evidence**: Studies show improved therapist responsiveness and reduced dropout rates when dashboards are integrated into care models (Fulmer et al., 2019).

Gamified Therapy Platforms

- **What they do**: Use features like progress bars, achievements, or reward systems to make therapeutic tasks feel more engaging and rewarding.
- **In this context**: Boost motivation by transforming exercises into manageable goals, particularly useful for younger clients or those struggling to stay invested.
- **Evidence**: Gamification has been linked to increased participation and task completion in digital mental health programs (Beg et al., 2024).

Practical Example: AI Chatbots Supporting Daily Engagement Between Sessions

AI chatbots are also being used to keep clients engaged between therapy sessions. These tools can help with homework assignments, mood tracking, and reinforcing key messages from therapy. In a pilot study of PracticePal, a chatbot paired with the Healthy Activity Program in rural India, clients with depression used the tool regularly for 29 out of 60 treatment days. Two-thirds of users engaged with more than half of the multimedia content. All participants completed treatment and showed significant improvement in depression symptoms. Counselors also said that clients followed between-session tasks more consistently than when using paper methods (Agrawal et al., 2025).

More broadly, research suggests that chatbots can improve follow-through by offering a private, nonjudgmental space for users to share their thoughts and feelings. Many are also available 24/7, which adds flexibility. A review of mental health chatbot apps found that users appreciated their constant availability and supportive tone. Some even said they felt more comfortable opening up to a chatbot than to friends or family. These qualities were linked to better engagement and follow-through. However, the review also noted risks, such as poor responses or unsafe handling of crisis situations (Haque & Rubya, 2023).

A new direction in chatbot design involves tools that reach out to users instead of waiting for them to start the conversation. One example is ComPeer, a generative peer-support chatbot tested in a one-week study with 24 participants. Compared to a traditional, user-initiated chatbot, ComPeer's proactive and context-aware messages led to more engagement and more sharing of feelings and life events. Users said the chatbot's consistent, peer-like tone and timely messages made it feel more supportive over time (Liu, Zhao, Liu, Wang, & Peng, 2024). This kind of proactive design could help reduce drop-off between sessions, keep skill practice on track, and support a stronger therapeutic connection in digital formats.

Concluding Remarks

Staying engaged between therapy sessions is often where progress falters. Without daily reinforcement, many clients lose momentum, skipping homework, falling back into old patterns, or disengaging entirely. This limits treatment impact, especially when clinicians cannot provide continuous support.

AI tools offer a practical way to extend care beyond the session. By delivering timely nudges, personalized tasks, and ongoing check-ins, these systems help clients stay focused on their goals between appointments. They do not replace human connection, but they do make follow-through more achievable.

When thoughtfully implemented, AI becomes a quiet but consistent companion to the therapeutic process, helping clients turn insights into habits, moments of reflection into progress, and weekly conversations into lasting change.

AI in Telehealth and Remote Psychological Practice

Telehealth has made mental health care more accessible, but virtual sessions pose unique clinical challenges. Therapists often struggle to read emotional cues or detect early signs of disengagement, and many clients find it harder to build rapport through a screen. These limitations can weaken the therapeutic connection and reduce session quality.

Opportunities and Benefits

AI-enhanced tools are beginning to close the gap between in-person and virtual care. These technologies support clinicians by revealing emotional cues that are often harder to detect through a screen. When integrated into telehealth platforms, AI can help clinicians maintain attunement, respond in real time, and personalize care more effectively—even when the session is remote.

Enhancing Emotional Attunement in Virtual Settings
AI can help restore emotional nuance in teletherapy by analyzing vocal tone, speech cadence, and facial microexpressions. These tools expose affective shifts that might otherwise be missed on screen, supporting deeper attunement and therapeutic presence.

Augmenting Therapist Awareness in Real Time
AI-powered dashboards provide clinicians with real-time indicators of emotional intensity, engagement, or relational friction. This allows therapists to adjust their pacing or approach mid-session, without diverting attention from the client.

Supporting Adaptive Intervention
By recognizing linguistic markers of distress, avoidance, or cognitive overload, AI systems can suggest tailored prompts or reframes drawn from evidence-based models. This helps clinicians respond with more precision, even when cues are subtle or ambiguous.

Improving Equity in Remote Session Quality
Clients who are less verbally expressive—or who rely more on nonverbal communication—may be disadvantaged in teletherapy. AI tools that interpret tone and facial cues can help reduce these gaps, helping to recognize emotional signals regardless of communication style.

Flagging Risk Indicators in the Moment
AI systems can detect patterns of withdrawal, agitation, or crisis—even when not clearly verbalized. These alerts support early, in-session intervention rather than delayed follow-up, especially in high-risk or emotionally volatile cases.

Example Applied AI Tools

A growing number of AI tools are being used to enhance virtual therapy sessions by making it easier to detect emotional nuance, tailor interventions, and respond in real time. The examples below illustrate ways AI has the potential to support in-session care delivery in telehealth environments.

AI-Enhanced Virtual Sessions
- **What they do:** Analyze vocal tone, facial expression, speech tempo, and language patterns in real time during video sessions.
- **Use in session**: Provide therapists with moment-to-moment insights into client affect, arousal, and engagement—helping illuminate signals that might otherwise be missed through a screen.
- **Evidence**: Studies report improved therapist awareness of emotional shifts and stronger rapport-building in remote sessions when AI tools are integrated (Psychology Today, 2024).

Augmented Clinical Dashboards
- **What they do:** Aggregate live session data—such as sentiment trends, verbal cues, and engagement markers—and highlight risk or rupture points in real time.
- **Use in session**: Help clinicians monitor emotional dynamics mid-session without breaking flow, flagging signs of distress, fatigue, or therapeutic rupture for timely response.
- **Evidence**: Research shows these systems improve detection of missed emotional cues and support more responsive interventions during virtual care (Mansoor & Ansari, 2024).

Session-Aware Conversational AIs
- **What they do**: Use natural language processing (NLP) to generate context-sensitive prompts, metaphors, or psychoeducation suggestions based on live client input.
- **Use in session**: Assist therapists in selecting timely, personalized interventions, especially when addressing avoidance, emotional withdrawal, or narrative disruption.
- **Evidence**: Pilot studies suggest that AI-suggested prompts enhance therapeutic alignment and increase client-reported session helpfulness in virtual settings (Habicht et al., 2025).

Practical Example: Using AI to Enhance Emotional Insight in Virtual Therapy

As mentioned earlier, one of telehealth's core limitations is the loss of in-room context—subtle emotional cues like tonal shifts, facial microexpressions, or brief hesitations that signal distress or disengagement. A 2024 pilot study by Mansoor and Ansari tested whether AI could help therapists recover some of that nuance in real time (Mansoor & Ansari, 2024).

In the study, clinicians conducted virtual sessions supported by an AI system that analyzed speech cadence, vocal tone, and language sentiment as the session unfolded. The system flagged potential emotional shifts or relational ruptures, offering in-the-moment insights without disrupting the session flow.

Therapists reported increased awareness of client affect and were more likely to pause, check in, or reframe when subtle shifts were flagged. Clients, in turn, rated sessions as more helpful and emotionally attuned. One therapist noted, "It was like having a second set of ears for things I usually notice in person but can't always catch over video."

Importantly, the AI did not interpret emotions or dictate action, but rather it amplified signals therapists might otherwise miss on screen. The result was not automation, but augmentation: a tool that helped preserve the depth of human connection in a medium that often flattens it.

Concluding Remarks

Telehealth has widened access to care, but at the cost of emotional visibility. In virtual sessions, key cues that guide empathy, pacing, and connection can be easily missed. AI tools are starting to fill that gap by sharpening therapists' awareness. By exposing subtle shifts in tone, expression, or engagement, AI helps clinicians stay attuned in real time. When thoughtfully integrated, these tools have the potential to preserve the human core of therapy, even when the room is digital.

Improving Diagnostic Precision

Mental health diagnosis remains one of the most difficult areas in clinical care. Many psychiatric conditions share overlapping symptoms that can vary significantly from person to person (Zhou, Zhao, & Zhang, 2022; Cruz-Gonzalez et al., 2025). Clinicians often rely on structured interviews and self-report tools, but these methods are limited by subjective interpretation, clinician experience, and inconsistencies in how symptoms are expressed across cultural or demographic lines (Lee et al., 2021; Day, Woldgabreal, & Butcher, 2022).

This diagnostic variability can contribute to increased rates of misdiagnosis and delays in appropriate treatment. A client with bipolar disorder might be misdiagnosed with unipolar depression and prescribed an ineffective—or even harmful—treatment plan. For complex or comorbid cases, the risk of error increases further. Improving diagnostic precision is essential—not just for choosing the right intervention, but for minimizing harm, reducing trial-and-error treatment cycles, and ensuring clients receive care that matches their needs (Babu & Joseph, 2024; Cruz-Gonzalez et al., 2025).

Opportunities and Benefits

AI is emerging as a valuable tool to improve diagnostic accuracy in mental health care. By analyzing large-scale data, identifying subtle symptom patterns, and applying consistent criteria, AI systems can help reduce the subjectivity that often leads to misdiagnosis. These technologies can enhance clinical judgment by surfacing distinctions that are difficult for humans to detect, especially in complex or overlapping cases. When trained on diverse populations and integrated thoughtfully, AI can support more precise, equitable, and efficient diagnostic processes.

Supporting More Precise Differential Diagnosis

AI models trained on large-scale clinical datasets may help distinguish between disorders with overlapping symptoms—such as bipolar vs. unipolar depression or PTSD vs. generalized anxiety—by detecting subtle patterns that can be difficult to identify in standard interviews.

Reducing Variability in Clinical Judgment

By applying consistent decision frameworks across patients, AI tools have the potential to reduce diagnostic differences stemming from individual clinician experience, theoretical bias, or inconsistent application of criteria.

Assisting with Complex or Comorbid Presentations

AI systems can analyze multiple symptom domains at once, which may help clinicians navigate diagnostic decisions in cases involving co-occurring conditions or atypical presentations.

Accounting for Cultural and Demographic Variability

When trained on inclusive and diverse datasets, AI tools could improve the accuracy of diagnoses across racial, cultural, or gender-diverse populations by recognizing variations in symptom expression that are often overlooked.

Reducing the Risk of Harm from Misdiagnosis

Improved diagnostic precision may reduce the likelihood of inappropriate or ineffective treatments—for instance, avoiding antidepressant monotherapy in undiagnosed bipolar disorder—which could help prevent symptom exacerbation and treatment delays.

Example Applied AI Tools

A range of AI technologies are being explored in clinical and research settings to support more accurate and consistent psychological diagnoses. These tools are intended to augment clinical expertise, helping clinicians navigate complex presentations, reduce diagnostic variability, and detect patterns that may be difficult to capture through traditional methods.

Deep Learning for Neuroimaging and EEG Analysis

- **What it does**: Applies machine learning models to functional magnetic resonance imaging (fMRI) and electroencephalogram (EEG) data to identify brain activity patterns associated with specific psychiatric conditions.
- **Potential benefit**: May support differentiation between disorders that share similar behavioral symptoms—such as distinguishing unipolar from bipolar depression—by surfacing underlying neural markers.
- **Supporting evidence**: Early studies suggest this approach could assist with complex cases and enable earlier identification of disorders like

schizophrenia or major mood disorders (Wang et al., 2022; Banerjee et al., 2019).

Natural Language Processing (NLP) for Voice and Text Analysis
- **What it does**: Analyzes features such as speech tone, word usage, and sentence structure from clinical conversations or self-reports.
- **Potential benefit**: May help surface linguistic cues associated with conditions like PTSD, depression, or psychosis, particularly in cases where symptoms are underreported or ambiguous.
- **Supporting evidence**: NLP tools have shown promise in detecting signs of psychological distress across diverse language inputs, with potential to support more consistent assessments (Banerjee et al., 2019).

Predictive Modeling with Electronic Health Records (EHRs)
- **What it does**: Uses AI to analyze EHR data—including diagnostic history, medications, comorbidities, and demographic factors—to flag patients at higher diagnostic or clinical risk.
- **Potential benefit**: Could reduce missed warning signs and help clinicians identify individuals who may warrant closer evaluation or differential diagnosis.
- **Supporting evidence:** Some models have demonstrated stronger predictive accuracy than standard screeners, particularly in identifying complex or high-risk cases (Wang et al., 2022).

Multimodal Fusion Models for Complex Presentations
- **What it does**: Integrates multiple data streams—such as facial expressions, speech patterns, and language cues—to build a richer profile of client symptoms.
- **Potential benefit**: May enhance diagnostic accuracy in cases with comorbid or overlapping presentations, where single-source assessments fall short.
- **Supporting evidence**: Research has shown that multimodal approaches outperform unimodal systems in identifying major depressive disorder and predicting symptom severity (Haque et al., 2019).

Practical Example: Multimodal AI to Reduce Diagnostic Variability in Depression

Accurate diagnosis of major depressive disorder (MDD) is notoriously difficult, due to its symptom overlap with other conditions. Traditional assessments often rely on structured interviews and self-report, which are influenced by clinician interpretation, cultural factors, and patient communication style, potentially leading to misdiagnosis and delayed treatment (Lee et al., 2021; Day, Woldgabreal, & Butcher, 2022).

To address this challenge, researchers at Stanford University developed a multimodal AI system designed to improve diagnostic precision for MDD by analyzing facial expressions, vocal tone, and language use simultaneously. The model was trained on the Distress Analysis Interview Corpus -Wizard of Oz (DAIC-WOZ) dataset, which includes clinical interview data from individuals with and without depressive symptoms (Haque et al., 2019).

Key Features of the Model:
- Speech analysis captured vocal prosody and timing patterns that often shift in depression.
- Facial data tracked subtle expressions, gaze, and head movement as nonverbal markers of affect.
- Linguistic analysis examined both the structure and content of responses, such as self-focus or negative language patterns.

Results:
- The model achieved over 83% sensitivity and specificity in identifying MDD.
- It also predicted Patient Health Questionnaire – 8 (PHQ-8) depression scores with an average error of just 3.67 points.
- Most importantly, combining all three modalities significantly outperformed any single input—suggesting that diagnostic accuracy improves when nonverbal and verbal cues are evaluated together (Haque et al., 2019).

Implications:
This study illustrates how AI might help standardize and enhance the diagnostic process, especially for complex presentations where subjective clinical judgment can vary. By illuminating subtle affective signals that may be missed in typical interviews, these tools could support more reliable identification of depression and reduce misdiagnosis driven by clinician bias or symptom ambiguity.

While further validation is needed in diverse, real-world settings, multimodal AI shows potential as a diagnostic decision support tool—one that augments clinician insight without replacing it.

Concluding Remarks

Improving diagnostic precision is one of the most pressing challenges in mental health care. While traditional methods rely on subjective tools that can miss subtle differences or misinterpret symptoms across diverse populations, AI technologies are beginning to offer support. From natural language processing to multimodal analysis, these systems show potential to augment clinician judgment by highlighting patterns that are easy to overlook, particularly in complex, overlapping, or culturally nuanced presentations. AI can serve as a second set of eyes: helping reduce misdiagnosis, improve treatment matching, and support more consistent, equitable clinical decision-making when used with care, oversight, and a commitment to clinical rigor.

Supporting Early Detection of Emerging Mental Health Concerns

Early signs of mental health concerns—such as mood shifts, disrupted sleep, or subtle social withdrawal—can appear weeks or months before a diagnosable condition takes hold. Yet these signals are often missed in traditional care settings, where screening relies on self-report, infrequent check-ins, or symptom escalation. Clients may not recognize these changes as meaningful or may hesitate to disclose them. As a result, opportunities for early, preventative support are often lost.

The absence of continuous, low-burden detection tools contributes to delayed intervention, especially in underserved or high-volume settings where proactive monitoring is not feasible. Strengthening early detection means identifying potential concerns before they meet diagnostic thresholds, opening the door for more timely support that could reduce the need for intensive care down the line (Zhu et al., 2020).

Opportunities and Benefits

AI tools are being explored as a way to support earlier detection of emerging mental health concerns—before they escalate into diagnosable conditions. By analyzing subtle patterns in language, speech, behavior, or health records, these systems may help reveal early indicators of emotional or cognitive change that may be missed in traditional screening.

When integrated into clinical or digital health settings, AI has the potential to function as a low-burden, continuous monitoring layer—identifying patterns that suggest elevated risk even in the absence of formal symptoms. This could be particularly valuable in time-limited, high-volume, or underserved environments where frequent clinician-led assessment is not feasible. Rather than replacing judgment, these tools can extend it—offering clinicians and clients more opportunities to intervene early and prevent worsening distress.

Detecting Subtle Early Changes
AI models can identify small shifts in language, speech, sleep, or digital behavior that may signal early emotional or cognitive distress—often before symptoms are consciously recognized or reported.

Enabling Continuous, Passive Monitoring
Rather than relying on occasional screenings, AI tools can analyze real-time data from smartphones, wearables, or journaling apps to monitor trends over time, offering early insight with minimal burden on users.

Flagging Risk in Resource-Limited Settings
In high-volume or low-access environments, AI systems can help prioritize attention by flagging individuals who show early risk markers—supporting earlier outreach even when clinician time is limited.

Prompting Earlier, More Targeted Interventions
By identifying elevated risk before conditions escalate, AI tools may help clinicians intervene earlier—potentially reducing the need for crisis care and improving long-term outcomes.

Reducing Reliance on Self-Report

Many clients underreport distress due to stigma, unawareness, or communication barriers. AI systems analyzing passive or behavioral data offer a complementary lens that does not depend on user disclosure.

Supporting More Equitable Detection

When trained on inclusive datasets, AI tools have the potential to detect emerging risk more consistently across cultural, racial, gender, and linguistic differences—helping reduce disparities in who is identified and when.

Example Applied AI Tools

Several AI tools are being tested to support earlier detection of mental health concerns, particularly in cases where symptoms are subtle, intermittent, or not yet clinically diagnosable. These tools focus on continuous, passive monitoring and pattern recognition, aiming to reveal early warning signs that might otherwise go unnoticed in routine care.

MoodCapture
- **What it does**: Uses a smartphone's front-facing camera to passively collect and analyze naturalistic facial images to detect depressive symptoms.
- **In this context**: Offers a low-burden, unobtrusive way to monitor mood in real-world settings, especially for individuals not currently engaged in care.
- **Evidence**: In a study involving 125,000 images, the tool showed predictive ability above chance in detecting depressive states, signaling potential for early detection (Nepal et al., 2024).

Ellipsis Health
- **What it does**: Uses voice analysis to detect early signs of anxiety and depression from short speech samples.
- **In this context**: Can be embedded into routine check-ins or digital health apps, offering early screening during casual conversations.
- **Evidence**: Early trials have demonstrated alignment with clinical measures of distress, showing promise for scalable early risk assessment (Zhang & Wang, 2024).

NLP-Based Journaling Apps
- **What they do**: Analyze written journal entries for linguistic markers of distress, such as self-focus, rumination, or negative emotion words.

- **In this context**: Provide early indicators of deteriorating mental health in users who may not verbalize concerns directly.
- **Evidence**: Studies show that linguistic cues can precede diagnosable symptoms by days or weeks, offering a window for early intervention (Banerjee et al., 2019).

Practical Example: Passive Monitoring for Emerging Depression Risk

A recent study explored the use of MoodCapture, an AI system designed to detect early signs of depression through passive image data collected from smartphone front-facing cameras (Nepal et al., 2024). The system analyzed over 125,000 naturalistic images from users with a history of major depressive disorder, examining facial expressions, environmental context, and lighting conditions to estimate depressive states.

Key Findings:
- The machine learning model showed predictive ability above chance in identifying depressive symptoms, even in the absence of user self-report.
- The system captured subtle emotional shifts in everyday environments, suggesting utility in detecting emerging risk prior to clinical deterioration.
- Researchers noted that the tool may be particularly helpful for younger individuals or those less likely to engage in traditional care, offering a non-intrusive path to earlier support.

Implications:
MoodCapture represents a class of AI tools that can enable low-effort, continuous monitoring, offering early signals when clinical thresholds have not yet been met. For systems facing capacity constraints, such tools may help prioritize outreach and extend support to users who are not actively seeking care. While further validation is needed, especially regarding ethical safeguards, this study highlights AI's potential to shift detection earlier in the trajectory of illness—closer to prevention than reaction.

Concluding Remarks

Early detection of mental health concerns remains a critical blind spot in current care models. Subclinical symptoms often emerge quietly and go unnoticed until they escalate into diagnosable conditions, missing key opportunities for early support. AI tools are beginning to show potential in closing this gap—not by diagnosing conditions, but by illuminating patterns that may signal emerging risk.

Through passive monitoring, pattern recognition, and low-burden engagement, these systems offer a new layer of insight that could complement traditional assessments. While more research is needed to ensure their reliability, safety, and equity, AI-based early detection tools may eventually help shift mental health care toward a more preventative, responsive model, potentially reaching people sooner, and reducing the burden of untreated illness over time.

Crisis Monitoring & Risk Detection

Mental health crises often emerge abruptly and escalate quickly, posing serious risks to safety and well-being. Unlike early-stage symptoms, which may unfold gradually, crisis states such as suicidal ideation, manic episodes, or acute psychosis can intensify significantly within hours or days. Traditional systems rely heavily on self-reporting or scheduled check-ins, which leaves significant blind spots between sessions. Clients may minimize distress, avoid disclosure, or be too overwhelmed to seek help. As a result, clinicians often lack real-time insight into worsening conditions, delaying intervention and increasing the risk of hospitalization or harm. Without tools for timely risk detection, mental health services remain largely reactive rather than preventive (Zhang & Wang, 2024).

Opportunities and Benefits

AI tools are increasingly being explored as a way to monitor real-time psychological risk, particularly in the critical moments between clinical encounters. By analyzing speech, behavior, or biometric data, these systems can reveal indicators of acute distress that might otherwise go unnoticed. Unlike traditional methods, which rely on periodic self-report or observation, AI offers the potential for continuous monitoring and faster escalation when needed.

While these tools are still under development, early research suggests they could help clinicians respond more quickly to emerging crises, prioritize high-risk cases, and extend monitoring into moments when clients are not actively engaged in care.

That said, below are some of the potential benefits AI technologies can provide for crisis monitoring and risk detection:

Detecting Crisis Signals Sooner
AI models can process changes in speech, language, or digital behavior that may indicate emotional dysregulation, suicidal ideation, or psychotic thinking—often days before a crisis is reported.

Enabling Real-Time Alerts
When risk thresholds are crossed, AI systems can generate timely notifications for clinicians or crisis teams, supporting earlier intervention and reducing response delays.

Bridging Gaps Between Sessions
Continuous or passive monitoring helps flag escalating distress between appointments, when risk may be rising but the client is not in contact with a provider.

Improving Risk Stratification
AI tools can help clinicians prioritize outreach and tailor responses by assigning risk scores based on individual behavior patterns, history, and symptom progression.

Reducing Reliance on Self-Disclosure
Clients in acute distress may be unwilling or unable to articulate what they are experiencing. AI-based systems provide an additional lens for identifying hidden or unspoken risk.

Scaling Crisis Detection in Underserved Contexts
In high-volume clinics or telehealth environments, AI tools can serve as a first-pass screening layer—helping to reach individuals at greatest need, even when human capacity is limited.

Example Applied AI Tools

AI-based systems are being developed and piloted to assist clinicians in identifying crisis-level mental health risks earlier and more accurately. These tools offer continuous monitoring, data-driven alerts, and additional insights in moments when clients are not actively seeking help or verbalizing their needs.

A Note on Shared Tools, Distinct Purposes

Some of the technologies used for crisis detection—such as voice analysis, wearable sensors, and sentiment analysis—also appear in the early detection strategies section. While the underlying tools may be similar, their application differs. In early detection, AI helps surface subtle signs before a condition becomes diagnosable. In crisis monitoring, those same tools are tuned to detect imminent risk and rapid deterioration. What shifts is the threshold for action, the time sensitivity, and the need for escalation. The following tools reflect that distinction, emphasizing real-time responsiveness and intervention readiness in high-risk moments.

Sentiment Analysis for Crisis Language

- **What it does**: Analyzes written or spoken language for signs of hopelessness, despair, or suicidal ideation.
- **In this context**: Can scan therapy transcripts, journal entries, or messaging platforms to flag critical language patterns.
- **Evidence**: In a study by Kuhail et al. (2024), sentiment-based AI models identified suicide risk with 35% greater accuracy than traditional clinical interviews.

Speech and Voice Pattern Detection

- **What it does:** Tracks vocal indicators—such as reduced variability, slowed cadence, or abrupt pauses—that can signal acute distress or dissociation.
- **In this context**: Used in telehealth or mobile apps to detect shifts in emotional state during conversations.
- **Evidence**: Beg et al. (2024) reported that voice-based AI tools show promise in identifying risk markers related to depressive and psychotic episodes in real time.

Wearable-Integrated Behavioral Monitoring

- **What it does**: Uses data from smartwatches or fitness trackers (e.g., sleep disruption, heart rate variability, reduced physical activity) to detect physiological indicators of emotional dysregulation.
- **In this context**: Helps monitor for warning signs of crisis in between appointments or outside of formal care.
- **Evidence**: Zhang & Wang (2024) describe how wearable-based models have been used to predict depressive relapse and agitation episodes with moderate accuracy.

AI-Driven Risk Scoring Platforms

- **What it does**: Aggregates data from electronic health records (EHRs), symptom logs, and real-time behavior to generate individualized crisis risk scores.
- **In this context**: Supports clinical decision-making by identifying high-priority patients for outreach.
- **Evidence**: Fulmer (2019) found that systems using predictive scores improved triage efficiency and reduced response times in behavioral health crisis teams.

Social Media Monitoring Systems

- **What it does**: Uses NLP and machine learning to scan public social media posts for patterns associated with suicidal ideation or emotional instability.
- **In this context**: May serve as an additional tool to identify high-risk individuals not connected to formal mental health care.
- **Evidence**: Babu & Joseph (2024) highlight how AI models have successfully flagged high-risk language patterns on social platforms, prompting timely intervention.

Practical Example: AI-Powered Suicide Risk Detection in Clinical Settings

One study conducted by Kuhail et al. (2024) evaluated the effectiveness of an AI-driven suicide prevention tool integrated into telehealth and clinical platforms. The tool used sentiment analysis, linguistic modeling, and engagement pattern tracking to detect signs of suicidal ideation in natural conversation.

Key Findings:

- **Higher Accuracy Than Clinicians**: The AI system identified suicide risk with 35% greater accuracy than traditional clinical interviews.
- **Earlier Detection**: On average, the model flagged indicators of suicidal ideation two weeks earlier than standard assessments.
- **Faster Intervention**: Clinicians responded to flagged cases 40% faster than in control groups, enabling more timely safety planning and reducing emergency hospitalization.

Implications:

This study highlights how AI can extend the reach of crisis response by monitoring for risk signals in real time—beyond what therapists can observe during scheduled sessions. By flagging subtle indicators of escalating distress, such tools can give providers a head start in preventing harm; however, ethical safeguards remain essential. Researchers emphasize the importance of clinician oversight, transparency in how alerts are used, and sensitivity to false positives that could undermine trust. Still, the findings suggest that when carefully implemented, AI can augment clinical care by detecting imminent risk earlier and improving crisis response workflows.

Concluding Remarks

As stated earlier, mental health crises can emerge suddenly and with devastating consequences—especially when early signs go unnoticed. Traditional systems, which rely on client disclosure or infrequent assessments, often fall short in detecting real-time danger. AI-powered monitoring tools are beginning to fill this gap by offering continuous, data-driven insights that may help identify crisis risk earlier and more accurately than conventional methods.

While these technologies are not a substitute for clinical judgment, they show promise as complements to care—flagging subtle warning signs in speech, behavior, or digital activity that might otherwise be missed. Real-time alerts, predictive scoring, and sentiment analysis can give providers a crucial window to intervene before harm occurs.

That said, these tools must be deployed with caution. Issues like false positives, consent, privacy, and clinical integration require careful consideration. Used ethically and under human oversight, AI can enhance the therapist's role in crisis care, helping shift the system from reactive response to proactive prevention.

AI in Child & Adolescent Psychology

Children and adolescents often face unique developmental challenges that can significantly impact their learning, emotional regulation, and social integration. These challenges, including speech and language difficulties, delayed diagnoses of developmental disorders, and difficulties with emotional regulation, are common in populations with neurodevelopmental conditions such as autism spectrum disorder (ASD), attention deficit-hyperactivity disorder (ADHD), and learning disabilities. Traditional methods of diagnosis, communication, and behavior management can be time-consuming and often leave gaps in early intervention, leading to missed opportunities for timely support.

Opportunities and Benefits

AI tools offer innovative solutions to directly address key challenges in child and adolescent psychology. For instance, by leveraging data-driven technologies, AI can enhance the early identification of developmental disorders, improve communication for children with speech and language challenges, and monitor emotional and behavioral patterns for early intervention. These technologies hold particular promise in supporting children in under-resourced settings by providing timely, personalized, and scalable interventions. In this section, we explore how AI is making significant strides in improving communication, diagnosing developmental conditions earlier, and offering proactive solutions for emotional and behavioral challenges.

Improved Communication for Children with Speech and Language Challenges

AI technologies, particularly speech recognition and translation tools, are improving accessibility for children with speech and language challenges. Tools like Google's Project Euphonia use machine learning to better understand diverse speech patterns, which can help children with neurological or speech-language disorders communicate more independently. These innovations foster more inclusive communication, allowing children to interact more effectively in both academic and social settings.

Early Identification and Diagnosis of Developmental Disorders

AI's role in early diagnosis is transformative, offering a way to detect developmental disorders like ASD, ADHD, and language delays at earlier stages than traditional methods. AI systems can analyze behavioral cues such as eye movement, vocal tone, and social interactions, significantly improving diagnostic accuracy and speed. Tools like Cognoa and SenseToKnow have demonstrated high accuracy in diagnosing children as young as 17 months, allowing for early interventions that are critical when neuroplasticity is at its peak.

Emotional and Behavioral Monitoring for Early Interventions

For children with emotional or behavioral challenges, especially those with ASD, AI tools such as wearable devices like the Q Sensor are proving invaluable. These devices monitor physiological markers of emotional arousal (e.g., electrodermal activity), allowing caregivers and educators to intervene before behaviors escalate.

Taking a proactive approach to monitoring emotional well-being not only has the potential to improve behavior management but also enhance the quality of life for children by addressing needs earlier, thereby reducing the likelihood of frequent and intensely escalating behaviors that interfere with functioning.

Example Applied AI Tools

The following tools illustrate how AI is being deployed to support assessment, intervention, and personalized learning in child and adolescent psychology. These technologies are shaping how professionals identify developmental concerns and deliver individualized support.

EndeavorRx (Akili Interactive)
- **What it does**: The first Food and Drug Administration (FDA)-approved video game treatment for ADHD that uses adaptive algorithms to improve attention function.
- **In this context**: Offers a non-pharmacological option for managing attention-related challenges in children, with built-in performance tracking.
- **Evidence**: Klimova and Pikhart (2025) discuss how gamified, AI-driven interventions can improve student engagement, emotional well-being, and academic outcomes, supporting the use of digital therapeutics in cognitive and attention training.

Cognoa
- **What it does**: AI-based autism diagnostic tool that analyzes video, audio, and parent questionnaire data to assess developmental behavior.
- **In this context**: Helps clinicians identify ASD earlier and streamline referrals for diagnostic follow-up.
- **Evidence**: Dawson et al. (2023) demonstrate the effectiveness of AI-enhanced behavioral assessments for early autism detection, achieving high sensitivity and specificity using mobile tools in children aged 17-36 months.

Project Euphonia (Google)

- **What it does**: An AI speech recognition initiative focused on improving voice interface accessibility for people with atypical speech patterns, such as those caused by neurological conditions or developmental disorders.
- **In this context**: Represents how AI can enhance inclusivity in digital communication by enabling better interaction between individuals with speech impairments and digital tools.
- **Evidence:** Klimova and Pikhart (2025) highlight the role of AI in improving communication and accessibility, noting that speech and language applications can contribute positively to student engagement and well-being, particularly for those with special needs.

Century Tech

- **What it does:** An AI-powered adaptive learning platform that uses cognitive science principles and real-time data to personalize instruction.
- **In this context**: Supports learners with difficulties by tailoring content complexity and pacing to individual needs, helping to increase engagement and reduce learning gaps.
- **Evidence**: Holmes et al. (2019) explain that AI-powered adaptive systems can deliver more inclusive and equitable education, especially when designed to support students with diverse learning profiles.

Q Sensor (Affectiva)

- **What it does**: A wearable device that measures electrodermal activity (EDA), providing real-time data on physiological arousal and stress.
- **In this context**: Used by educators and mental health professionals to track emotional responses and anticipate behavioral challenges in students with ASD.
- **Evidence**: Ferguson et al. (2019) found that changes in EDA, recorded via Q Sensor, were linked to behavioral incidents, offering a potential tool for early intervention.

Practical Example: AI-Assisted Early Autism Detection in Toddlers

A 2023 study by Perochon et al. evaluated the SenseToKnow mobile application, which uses AI to analyze behavioral and speech patterns in toddlers for early detection of autism spectrum disorder (ASD). The tool was administered during pediatric well-child visits, and it analyzed video-recorded responses to social stimuli using computer vision and machine learning.

Key Findings
- **Strong Diagnostic Accuracy**: The AI model achieved 87.8% sensitivity and 80.8% specificity in identifying ASD among children aged 17–36 months, with an area under the curve (AUC) of 0.90. When combined with the traditional Modified Checklist for Autism in Toddlers-Revised with Follow-up (M-CHAT-R/F) questionnaire, AUC increased to 0.97.
- **Scalable and Efficient**: The digital tool was completed in less than 10 minutes and proved feasible for use in real-world clinical settings, including primary care visits.
- **Improved Equity**: The model's performance remained consistent across subgroups defined by sex, race, and ethnicity—addressing disparities typically observed with traditional screening tools.
- **Personalized Feedback**: The system provided interpretable, child-specific insights based on behavioral phenotypes, enabling targeted referrals and early intervention planning.
- **Limitations Noted:** While the tool shows promise, the study emphasized that AI-assisted screening should supplement—not replace—comprehensive clinical evaluations and must be implemented with attention to potential sampling biases and model transparency.

This example highlights how digital phenotyping powered by AI can make early autism screening more accurate, equitable, and scalable, especially in settings where traditional assessments may fall short.

Concluding Remarks

AI-driven tools are making significant strides in child and adolescent psychology, offering promising solutions to address developmental challenges and enhance intervention strategies. By enabling earlier identification of developmental disorders, providing personalized learning support, and offering real-time emotional and behavioral monitoring, AI has the potential to transform how clinicians, educators, and caregivers support children. These technologies can particularly benefit children in under-resourced environments by improving access to quality care and streamlining diagnostic and intervention processes.

Reducing Bias and Variability in Psychological Assessment

Psychological assessments play a key role in diagnosis and treatment planning, but they can be affected by bias and variability. Studies suggest that clinician judgment may be influenced—often unconsciously—by factors such as race, gender, or cultural background, potentially contributing to disparities in care (Babu & Joseph, 2024; Zhang & Wang, 2024). In addition, self-report measures, while useful, may be limited by recall inaccuracies or social desirability effects. AI presents one possible avenue for addressing these challenges by introducing data-driven tools that aim to support greater consistency, reduce subjectivity, and promote more equitable assessment practices (Kuhail et al., 2024).

Opportunities and Benefits

Psychological assessments are foundational to diagnosis and treatment, but they can be far from neutral. Clinician interpretation can be shaped by unconscious bias, cultural blind spots, and inconsistent protocols. AI tools may support more standardized assessments, help reduce certain types of human bias and contribute to more equitable evaluations. This can be helpful particularly for populations that have historically been misdiagnosed or underserved. Early research suggests these systems could complement clinician expertise in real time, especially in high-volume or resource-limited settings.

Note: The reader is referred to the AI in Neuropsychology section of Chapter 8: AI in Health Psychology for innovations involving cognitive and neuropsychological assessments.

Standardizing Clinical Judgment
AI-powered tools analyze language, tone, and behavioral data through consistent algorithms, reducing the variability that often comes with human interpretation. These systems apply the same criteria across cases, helping minimize diagnostic drift and bringing structure to subjective processes.

Reducing Racial, Cultural, and Gender Bias
When trained on inclusive datasets, AI models can better detect culturally specific

expressions of distress and avoid over-relying on norms grounded in majority populations. This improves the fairness of assessments for racial and ethnic minorities, women, and LGBTQ+ individuals—groups historically underserved by traditional tools.

Enhancing Objectivity in Emotional Evaluation

AI-driven facial recognition and speech analysis can detect micro-expressions, tone shifts, or verbal cues in real time. These data points provide clinicians with unbiased indicators of emotional state, complementing (and at times correcting) human perception.

Cross-Validating Self-Report Data

AI can combine self-reports with passive data inputs—such as voice, typing behavior, or facial expression—to create a more complete, reliable picture of mental health. This is especially helpful when memory gaps or social desirability skew self-disclosure.

Supporting Early Detection in Nontraditional Presentations

AI systems excel at identifying patterns that humans may miss, especially when symptoms do not match textbook cases. By analyzing subtle language shifts or behavior markers, these tools can flag early signs of anxiety, depression, or trauma, improving the odds of timely and accurate intervention.

Improving Consistency in Risk Assessment

Machine learning models trained on diverse patient data can flag suicide risk, emotional instability, or symptom escalation based on structured patterns. These tools reduce reliance on clinician gut instinct, helping prevent underdiagnosis in high-risk populations.

Structuring Assessment Protocols Across Settings

AI platforms often guide clinicians through standardized workflows, ensuring that critical diagnostic steps are followed regardless of setting or provider experience. This reduces omissions due to time constraints or individual habits, especially in high-volume or under-resourced clinics.

Used thoughtfully, AI in assessment is not about replacing human expertise; it is about backing it up with data. When designed for fairness and deployed with care, these tools can help push mental health diagnostics toward a future that is more consistent, inclusive, and evidence-based.

Example Applied AI Tools

Several AI tools are being explored to help reduce subjectivity in psychological evaluation. These technologies use data-driven methods to detect symptoms, assess emotional cues, and mitigate the influence of clinician bias.

Natural Language Processing (NLP) for Diagnostic Support

- **What it does**: NLP analyzes speech content, tone, and sentiment during clinical interviews or therapy sessions.
- **In this context**: Helps identify indicators of depression, anxiety, or mood disorders by interpreting language use patterns that may be overlooked by clinicians.
- **Evidence**: In Kuhail et al. (2024), clinicians could not reliably distinguish AI-generated responses from human therapeutic responses, suggesting NLP-based tools can meet clinical standards for empathy and accuracy.

Facial Recognition and Microexpression Tracking

- **What it does**: AI software detects facial expressions, micro-movements, and nonverbal signals to assess emotional states.
- **In this context:** Provides objective, real-time feedback about a patient's affect, reducing reliance on clinician interpretation alone.
- **Evidence**: Beg et al. (2024) describe the use of computer vision to detect emotional dysregulation, supporting more accurate assessment in mood disorders.

Algorithmic Risk Assessment Tools

- **What it does**: AI models process diverse patient data—such as history, behavior, or language—to flag suicide risk or psychological distress.
- **In this context**: Offers early detection capabilities and reduces dependence on subjective clinical judgment.
- **Evidence**: Babu & Joseph (2024) highlight how AI models can accurately detect risk levels in high-stakes environments like emergency mental health.

Bias Mitigation Through Inclusive Training Data

- **What it does**: AI tools are trained on datasets that reflect gender, cultural, and socioeconomic diversity.

- **In this context**: Reduces demographic bias by making AI evaluations more representative across populations.
- **Evidence**: Fulmer (2019) emphasizes that chatbot-based interventions trained on diverse populations are more effective and equitable in cross-cultural settings.

Practical Example: AI-Driven Decision-Support Systems in Mental Health

A study conducted by Kuhail et al. (2024) examined the effectiveness of AI-assisted psychological assessments compared to traditional clinician-led evaluations. The study assessed the diagnostic accuracy, inter-rater reliability, and fairness of AI-driven tools across a diverse clinical population. The findings revealed:

- **Increased Diagnostic Consistency**: AI-powered assessments improved inter-rater reliability, reducing variability in clinician diagnoses and ensuring greater diagnostic standardization.
- **Reduction in Implicit Bias**: AI-driven tools identified mental health conditions at similar rates across demographic groups, whereas human clinicians exhibited subtle biases in diagnostic judgments.
- **Improved Early Detection**: AI-assisted assessments detected early symptoms of generalized anxiety disorder and major depressive disorder with greater accuracy than traditional diagnostic methods.

These findings suggest that AI-powered psychological assessment tools have the potential to enhance diagnostic reliability and reduce disparities in mental health evaluations, leading to more equitable treatment outcomes.

Concluding Remarks

As AI becomes more involved in psychological assessment, early findings suggest it may help reduce certain types of diagnostic bias and bring more consistency to evaluations. Tools that analyze language and behavior, for example, show potential for creating more standardized assessments across different populations.

At the same time, adding AI to clinical workflows brings up important concerns about transparency, fairness, and possible unintended effects. To use these tools responsibly, there will need to be ongoing research, strong ethical oversight, and input from a wide range of communities.

Improving Treatment Selection and Planning with AI

Selecting the right psychological treatment is often a slow, uncertain process. Many clinicians still rely on trial-and-error methods, testing out different approaches, medications, or therapeutic modalities until something works. This is partly because individuals with the same diagnosis can respond very differently to the same treatment. Despite decades of research, there are few tools that help clinicians predict which interventions will be most effective for a particular patient. As a result, clients may face delays in symptom relief, reduced motivation, or premature dropout.

These challenges are compounded by the complexity of clinical data, ranging from comorbidities and prior treatment responses to patient preferences and behavioral trends. Without effective tools to synthesize and interpret this information, it is difficult to make timely, personalized care decisions that fully reflect the needs of each individual.

Opportunities and Benefits

AI is being explored as a tool to support more informed and personalized treatment selection. By analyzing large datasets and identifying patterns across clinical histories, behavioral indicators, and treatment outcomes, AI systems may offer clinicians new insights into what is likely to work for whom. Rather than replacing professional judgment, these technologies are designed to augment it, helping to reduce guesswork, improve responsiveness, and make care more efficient and tailored.

Here are ways AI might support clinicians in navigating the challenges of treatment planning:

Supporting Smarter Matching Between Patients and Treatments
AI models trained on large, diverse clinical datasets can help reveal patterns linking individual profiles with effective interventions—such as CBT, DBT, or specific

medications. These systems may help reduce reliance on trial-and-error planning by offering data-informed starting points (Torous et al., 2021).

Identifying Early Signs of Treatment Misfit
Rather than waiting for symptoms to worsen or progress to stall, AI systems can monitor early behavioral or engagement patterns that might suggest a treatment is not working as intended. This may help clinicians pivot sooner and reduce time spent on less effective approaches (Thakkar, Gupta, & De Sousa, 2024).

Extending Monitoring Between Sessions
AI tools integrated with mobile apps, wearables, or digital journals can track passive indicators—like mood, sleep, activity, or speech tone—offering clinicians a fuller picture of patient experience over time. This can enable more adaptive care and quicker follow-up when needed.

Enhancing Engagement Through Adaptive Tools
Some AI-enabled platforms, such as chatbot-guided therapy tools, offer personalized content and encouragement between sessions. These tools can reinforce treatment principles, increase adherence, and surface potential issues that might otherwise go unreported (Habicht et al., 2025).

Bringing Underused Patient Data into the Process
Behavioral trends, therapy transcripts, and clinical history often go unanalyzed in real-time decision-making. AI tools can synthesize these disparate data streams to highlight relevant insights that may support more tailored planning—especially in complex or comorbid cases (Zheng et al., 2020).

Helping Scale Care in High-Demand Settings
In overburdened systems, AI-assisted platforms may help standardize parts of the treatment planning process—supporting care decisions when clinician time is limited. This could be particularly useful in rural or low-resource contexts, where access to specialty care is often delayed (Zhang et al., 2025).

Example Applied AI Tools

AI is not just theoretical anymore; it is already being tested in real-world mental health settings. The tools listed below represent some of the technologies showing early promise in helping clinicians personalize treatment more effectively, offering an extra layer of insight when used thoughtfully.

Deep Learning Models for Treatment Prediction

Deep learning systems are being used to analyze clinical neuroimaging and genetic data to forecast how patients might respond to specific medications or therapies. These models can reveal patterns in brain connectivity, neurotransmitter activity, or behavioral markers, helping clinicians choose interventions with a higher chance of success (Wang et al., 2022).

Natural Language Processing (NLP) in Clinical Conversations

As described in the Improving Diagnostic Precision section of this chapter, NLP tools can detect cognitive distortions, emotional shifts, or disengagement by analyzing clinical conversations. In treatment planning, these insights can help clinicians adjust care dynamically based on language cues during or between sessions. For example, automated analysis of free speech has been shown to predict the onset of psychosis in high-risk youth with significant accuracy (Bedi et al., 2015). Broader reviews have also confirmed NLP's role in detecting depression and suicide risk by analyzing vocal and textual patterns (Cummins et al., 2015).

Electronic Health Record (EHR) Pattern Recognition

Building on the predictive models that were discussed in earlier sections of this chapter, EHR-based tools can also support smarter treatment matching. By surfacing underused trends—like past medication responses or comorbidities—AI can help clinicians make more personalized decisions based on a fuller picture of the patient's history. For example, deep learning models trained on EHR data have been used to flag patients at high risk for suicide attempts, enabling earlier intervention and tailored care planning (Zheng et al., 2020).

Wearables and Real-Time Monitoring

AI systems are starting to use more than just what patients say—they are watching how they say it. Tools that analyze subtle changes in facial expression and vocal tone are showing promise in identifying mental health shifts that might otherwise go unnoticed. In a large clinical study, researchers used automated facial image analysis and audio signal processing to detect depression based on facial movements and vocal prosody during therapy sessions. These nonverbal cues—measured with computer vision and machine learning—mapped onto clinical depression ratings with up to 88% accuracy (Cohn et al., 2009). This kind of multimodal monitoring, if integrated into digital tools or wearables, could offer clinicians earlier warning signs and a richer picture of patient status without requiring additional effort from the patient.

When brought together, these tools can give mental health professionals a more complete and data-informed picture of what is working, what is not, and where to go next.

Practical Example: AI-Supported CBT in Real-World Practice

A UK-based observational study (Habicht et al., 2025) examined the integration of an AI-enabled therapy support tool into group-based cognitive behavioral therapy (CBT) programs delivered through the National Health Service (NHS) Talking Therapies. The goal was to improve engagement with therapeutic exercises between sessions—a common challenge in CBT that can contributes to poor outcomes and high dropout rates. The study included 244 patients with depression or anxiety, some of whom used the AI tool alongside standard group therapy, while others followed traditional CBT methods without AI support.

AI Technology Applied

The tool, powered by a generative AI language model Generative Pretrained Transformer-4 (GPT-4), functioned as a conversational agent that guided patients through therapist-assigned CBT materials between sessions. It used natural language prompts to deliver exercises, provide psychoeducation, and support reflective thinking. The system was embedded with guardrails and supervised AI layers to ensure clinical safety, and it personalized content based on user input. Unlike static workbooks, the tool offered a responsive, interactive experience that encouraged patients to stay engaged outside of therapy sessions.

Key Findings

Patients who used the AI-enabled support tool attended more sessions, missed fewer appointments, and were significantly less likely to drop out of treatment. Clinical outcomes also improved: the AI group saw higher rates of reliable improvement, recovery, and reliable recovery compared to the control group. Importantly, the level of engagement with the AI tool was positively correlated with both adherence and treatment success—suggesting that the personalization and real-time interaction it provided played a meaningful role in enhancing therapeutic outcomes (Habicht et al., 2025).

Concluding Remarks

This section highlighted how AI is beginning to support treatment planning in counseling and clinical psychology by offering additional insight that can inform decision-making. When applied thoughtfully, early research suggests AI can help reduce delays in finding effective care, enhance patient engagement, and enable more personalized, data-informed strategies; however, these tools are still evolving. Realizing their full potential will depend on continued research, diverse and representative datasets, and clear clinical guidelines. The promise is there, but responsible implementation will determine its impact.

Reducing Administrative Burden in Mental Health Practice

Administrative work is consuming a growing share of clinicians' time. From documenting therapy sessions and managing billing codes to ensuring compliance with regulatory standards, mental health professionals often spend up to 40% of their workweek on nonclinical tasks (Lee et al., 2021; Beg et al., 2024). While essential for care continuity and accountability, these responsibilities can limit availability for client interaction, extend work hours, and increase the risk of burnout.

This burden is especially acute in high-volume or public-sector settings, where clinicians face tight schedules and limited support. Manual documentation processes also increase the risk of inconsistencies, errors, and missed reimbursement, compounding stress and reducing job satisfaction. For many providers, administrative strain undermines both care quality and clinician well-being—posing a challenge to the sustainability of mental health services.

Opportunities and Benefits

AI technologies are emerging as practical tools to help reduce the documentation and administrative workload facing mental health professionals. By automating routine tasks—such as note-taking, transcription, compliance checking, and scheduling—these tools can free up time, reduce after-hours work, and ease cognitive strain. While they are not a replacement for clinical expertise, AI systems can serve as behind-the-scenes support, helping clinicians focus more on care delivery and less on paperwork. Below are several ways these technologies are being used or explored to address administrative burden in mental health settings:

Time Savings Through Automated Documentation

AI transcription tools—like Nuance's Ambient Clinical Intelligence (ACI) or Eleos Health—can generate clinical notes in real time, reducing the hours clinicians spend on post-session paperwork. Some platforms report cutting documentation time by up to 40%, freeing up more hours for client care and personal recovery.

Less Work After-Hours and More "Pajama Time"

By capturing notes during or immediately after sessions, AI tools help reduce the need for evening or weekend catch-up work. Clinicians report improved work-life balance and less mental fatigue when administrative tasks are handled during the day.

Improved Accuracy and Regulatory Compliance

AI documentation platforms can flag incomplete entries, missing billing codes, or noncompliant phrasing. This helps reduce audit risk, avoid billing denials, and maintain accurate, standardized records that meet insurance and legal requirements.

Reduced Cognitive Load

AI takes on the "mental overhead" of formatting, categorizing, and repeating standard phrases—allowing clinicians to focus on higher-value thinking. This can improve focus during sessions and reduce burnout over time.

Real-Time EHR Integration

When linked to electronic health records (EHRs), AI-generated notes sync automatically, cutting down on duplicate entry and ensuring consistency across multidisciplinary teams. This is especially valuable in high-volume or collaborative care settings.

Scalable Support in Under-Resourced Systems

In clinics with long waitlists and limited staffing, AI can help keep up with demand—handling tasks like appointment scheduling, session summaries, or documentation prep. This enables clinicians to serve more clients without compromising quality.

Better Focus During Sessions

When clinicians do not have to multitask by taking notes or remembering billing codes mid-session, they can be more present with their clients. This can strengthen the therapeutic alliance and improve session quality.

Example Applied AI Tools

A growing set of AI-powered platforms is being used to support the documentation and administrative demands of mental health professionals. These tools aim to reduce time spent on routine tasks, ensure regulatory compliance, and streamline workflows—especially in busy or resource-limited environments. Below are several notable examples:

Eleos Health
- **What it does**: Uses natural language processing (NLP) to transcribe therapy sessions in real time, extract clinical insights, and generate structured progress notes.
- **In this context**: Helps clinicians complete documentation more quickly and consistently, while maintaining quality and compliance standards.
- **Evidence**: A randomized trial showed that therapists using manual methods took an average of 69 hours to complete their notes, while those using Eleos completed theirs in 14 hours, 55 hours faster on average. This improvement in efficiency was accompanied by enhanced documentation quality and reduced burnout (Sadeh-Sharvit et al., 2023).

Suki AI
- **What it does**: A voice-enabled digital assistant that allows clinicians to dictate notes, which are transcribed and structured using AI.
- **In this context**: Designed to streamline EHR documentation through speech recognition, smart templates, and predictive text.
- **Evidence**: Clinicians reported reduced after-hours work and improved efficiency, saving an average of five hours per month on unscheduled documentation tasks (Peterson Health Technology Institute, 2025).

Augmedix
- **What it does**: Combines live medical scribes with AI-powered transcription to generate EHR-ready documentation during or after sessions.
- **In this context**: Ideal for high-volume settings where clinicians need real-time support without sacrificing detail or accuracy.
- **Evidence**: While mental health-specific trials are ongoing, Augmedix has demonstrated improved workflow efficiency and clinician satisfaction in other clinical environments (Cruz-Gonzalez et al., 2025).

Ambient Clinical Intelligence (ACI)

- **What it does**: Developed by Nuance, ACI uses ambient voice capture and NLP to create clinical documentation from natural dialogue during sessions.
- **In this context**: Enables clinicians to remain fully engaged with clients without pausing to take notes.
- **Evidence**: ACI users reported up to 40% reductions in documentation time, with increased session presence and fewer errors (Kuhail et al., 2024).

AI-Powered Compliance Checkers

- **What they do**: Scan documentation for billing, legal, or regulatory compliance—flagging missing codes or incomplete entries.
- **In this context**: Reduce the risk of rejected claims, missed reimbursement, and audit-related stress.
- **Evidence:** Tools cited by Zhang & Wang (2024) improved documentation accuracy and reimbursement rates in behavioral health settings.

Journal Summarization and Pattern Detection Tools

- **What they do**: Tools like Relationship Fitness for Men enable clients to journal and share reflections between sessions; in development are AI layers that summarize key themes for clinicians.
- **In this context**: Help reduce the time clinicians spend reviewing lengthy content while discovering clinically meaningful patterns.
- **Evidence**: Siddals et al. (2024) found these types of tools can reduce admin load, improve client engagement, and make between-session insights more clinically actionable.

Practical Example: Implementing Voa—An AI Documentation Tool to Reduce Administrative Load

To help reduce documentation demands, a Brazilian public health system piloted Voa, an AI-powered clinical tool designed to streamline paperwork while preserving quality and compliance.

Technology Applied

Voa uses a combination of speech recognition and natural language processing (NLP) to generate structured clinical notes from live or recorded therapy sessions. It extracts clinically relevant content and auto-populates it into standardized documentation templates that align with Brazilian electronic health record (EHR) requirements.

Study Design and Implementation

The tool was deployed in primary care clinics across Brazil's public healthcare system. Mental health clinicians used Voa during their routine sessions, and outcomes were evaluated based on documentation time, note quality, and clinician satisfaction (Basei de Paula et al., 2025).

Key Findings

- **Reduced Documentation Time:** Clinicians using Voa completed documentation up to 44% faster than those using manual methods, freeing time for more client appointments.
- **Improved Record Quality:** Over 80% of notes generated with Voa met national auditing standards for completeness and compliance.
- **Higher Clinician Satisfaction:** Therapists reported a reduction in end-of-day paperwork, improved workflow efficiency, and better ability to stay engaged during sessions.
- **Sustained Clinical Autonomy:** Although Voa suggested summaries, clinicians retained final control, ensuring human oversight and professional discretion in documentation.

Implications

This example illustrates how AI tools like Voa can help reduce administrative strain in everyday practice. By streamlining note generation and supporting regulatory compliance, these technologies create more room for what clinicians are trained to do: deliver care. Especially in high-volume public health systems, tools like Voa offer a pathway to more sustainable, patient-centered practice.

Caveats

While promising, adoption of AI administrative technologies must be guided by careful attention to data privacy, contextual accuracy, and clinician autonomy. These tools should enhance—not replace—professional judgment, and their integration must be accompanied by training and clear clinical protocols.

Concluding Remarks

Administrative demands are a growing source of strain in mental health care, pulling time and energy away from clinical work. AI tools offer a practical way to lighten this load by automating routine documentation tasks, improving compliance, and streamlining workflows. While these technologies are not a fix-all—and must be implemented with strong ethical and privacy safeguards; they are beginning to free up clinician capacity in meaningful ways. When used responsibly, AI can support a more sustainable model of care, where mental health professionals spend less time on paperwork and more time with the people they're here to help.

Conclusion

AI is not a fix-all for the challenges facing mental health care, but it is beginning to offer practical support where it is needed most. From reducing administrative load to sustaining engagement between sessions, AI tools are helping clinicians do their jobs more effectively and clients stay connected to care more consistently.

As this chapter has shown, AI is being integrated into counseling and clinical psychology workflows. It offers ways to reinforce therapeutic goals, bridge between-session gaps, reduce dropout rates, support diagnoses, aid in treatment planning, and enhance earlier detection and crisis prevention—areas where precision and timeliness can make a life-changing difference.

We must not forget that these tools come with real risks: over-reliance, bias, misinterpretation, and ethical concerns. The challenge ahead is not just about developing better algorithms, but rather it is about ensuring that AI serves as a supplement, not a substitute, for clinician judgment and the therapeutic relationship. Done right, AI can free up time, reduce burnout, and expand access. This can help mental health professionals focus on what matters most: the people they serve.

In the next chapter, we shift from therapy rooms and clinics to the courtroom. Forensic psychologists are beginning to explore how AI might support risk assessment, case analysis, and legal decision-making. But the stakes are even higher—and the ethical questions more complex. We'll examine where AI is showing potential in forensic contexts, and what must be done to ensure that these systems promote justice, not bias.

Chapter 5: AI in Forensic Practice

Forensic psychologists play a vital role in the justice system by conducting evaluations related to risk, competency, and credibility—assessments that can significantly influence sentencing, release decisions, and other legal outcomes. Yet many aspects of forensic assessment remain resource-intensive and subject to variability across practitioners. As caseloads grow and concerns about fairness and consistency intensify, interest is increasing in approaches that are both scalable and evidence-informed.

Artificial intelligence (AI) is beginning to show potential in supporting these efforts. Emerging tools such as machine learning, natural language processing (NLP), and behavioral analytics are being explored to analyze courtroom speech, identify patterns in forensic interviews, and assess risk factors. These technologies are not designed to replace expert judgment but rather to enhance it—helping improve consistency and drawing attention to areas that warrant deeper clinical review.

This chapter examines how AI is currently being tested and applied within key areas of forensic psychology. It also considers the ethical, legal, and practical challenges of introducing these technologies in high-stakes contexts. While early findings suggest AI could improve efficiency and help reduce certain biases, its impact will ultimately depend on thoughtful integration, rigorous validation, and clear boundaries around its role in legal decision-making.

Evaluating Competency to Stand Trial

Determining a defendant's competence to stand trial (CST) is one central task in forensic psychology, requiring the evaluation of whether an individual can understand legal proceedings and effectively participate in their defense (Tortora et al., 2020). These assessments are crucial for upholding due process, but they can be time-consuming, dependent on clinical interpretation, and subject to significant variation across evaluators. Despite the use of structured tools and interviews, detecting subtle cognitive impairments that may affect legal comprehension remains a challenge (Hogan et al., 2021). In addition, the increasing number of cases often leads to delays in evaluations and can result in inconsistencies in the findings.

Opportunities and Benefits

AI has the potential to significantly improve competency to stand trial (CST) evaluations by addressing the challenges of inconsistency, inefficiency, and delayed assessments. By leveraging technologies such as natural language processing (NLP), machine learning, and behavioral analytics, AI can help streamline the evaluation process, making it faster and more consistent. These tools can analyze cognitive and linguistic patterns, detect subtle impairments, and ensure that every case is evaluated with the same rigorous standards, regardless of the evaluator. AI systems can also prioritize cases based on risk, ensuring that evaluations in high-volume jurisdictions are conducted more efficiently. When implemented responsibly, AI has the potential to reduce evaluator bias, improve detection of cognitive and psychiatric impairments, and support more evidence-based decisions in determining CST.

Faster Evaluations Without Sacrificing Quality
AI tools like CogEval AI and CompBot automate time-consuming tasks—like scoring tests, transcribing interviews, and flagging risk factors—cutting turnaround time by up to 40% (Ogunwale et al., 2024). This is crucial in jurisdictions facing long backlogs and excessive pretrial detention.

More Consistent, Less Subjective Assessments
CST evaluations depend heavily on clinical judgment, which can vary. AI systems apply the same standards to every case, analyzing speech, behavior, and cognition using consistent models. Studies show NLP tools can match human evaluators

91% of the time while flagging issues like disorganized thinking more systematically (Starke et al., 2023).

Improved Detection of Cognitive and Psychiatric Impairment
Not all impairments are obvious. AI-enhanced platforms like NeuroAI Insight help surface subtle cognitive issues—like attention deficits or executive functioning deficits—that may impact legal comprehension. In pilot studies, machine learning models classified competency with 85–90% accuracy, even in complex cases (Lee et al., 2021).

Smarter Triage in High-Volume Systems
With clinician shortages and overloaded dockets, prioritizing evaluations matters. Tools like CompBot offer risk scores based on behavior, history, and neurocognitive patterns, helping forensic teams decide who needs a full workup and who may not—without compromising fairness (Hogan et al., 2021).

Enhanced Legal Relevance Through Simulated Testing
AI-driven simulations—like the Forensic Interview Simulation platform—offer new ways to assess legal comprehension. These tools pose standardized questions through virtual avatars and analyze responses for coherence and reasoning. This creates measurable indicators of competence beyond just interview impressions (Zhang & Wang, 2024).

Reduced Impact of Evaluator Bias
AI does not eliminate bias on its own, but when trained on diverse datasets and used with clinical oversight, it can help reduce disparities in how CST is determined across racial and cultural lines. Tools that quantify reasoning, memory, and comprehension create a more level playing field, especially when human judgment is vulnerable to implicit bias.

Competency evaluations will not be fully automated, nor should they. But with responsible integration, AI can help evaluators move faster, work more consistently, and back their decisions with clearer, data-informed evidence. The right tools do not replace forensic psychologists; however, they have the potential to make their work sharper, faster, and better supported—especially when the legal system depends on getting it right.

Example AI Tools

A growing number of AI tools are being developed to assist with competence-to-stand-trial (CST) evaluations. While these technologies are still in early stages, they offer promising support for streamlining assessments, improving consistency, and augmenting clinical judgment with additional data inputs.

CogEval AI – Cognitive Screening and Decision Support Tool

- **What it does**: Uses machine learning algorithms to evaluate memory, executive functioning, and attention span based on responses to standardized tests and natural language prompts.
- **In this context**: Helps forensic evaluators detect cognitive deficits that may indicate incompetence, especially in complex cases involving brain injury or developmental disorders.
- **Evidence**: In preliminary validation studies, CogEval AI achieved 88% accuracy in classifying competency status compared to board-certified forensic evaluators (Lee et al., 2021).

NLP-Based Legal Comprehension Analyzer

- **What it does**: Applies Natural Language Processing (NLP) to assess a defendant's understanding of legal terminology, courtroom roles, and the consequences of plea decisions.
- **In this context**: Provides quantifiable scores for verbal coherence, logical reasoning, and legal concept comprehension during CST interviews.
- **Evidence**: Starke et al. (2023) report that this tool reliably identified impaired reasoning and confusion in simulated interviews, matching human judgment with 91% agreement.

AI-Supported Forensic Interview Simulation (AFIS)

- **What it does**: Delivers virtual interview simulations using AI-powered avatars that ask standardized legal questions and analyze the quality of responses.
- **In this context:** Evaluates a defendant's ability to participate in legal defense through scenario-based questioning, reducing variability in interviewer technique.
- **Evidence**: Zhang & Wang (2024) report that AFIS improved consistency in CST determinations across examiners in a pilot study of 60 cases.

NeuroAI Insight – Neuroimaging Analysis Platform

- **What it does**: Integrates functional magnetic resonance imaging (fMRI), computer tomography (CT), and electroencephalogram (EEG) data with AI algorithms to identify structural or functional brain anomalies associated with psychiatric or neurological disorders.

- **In this context**: Supports assessments of competence by providing objective evidence of brain dysfunction that may impair rational understanding or decision-making.
- **Evidence**: Tortora et al. (2020) found that NeuroAI Insight improved the detection of frontal lobe damage and correlated impairments in CST performance.

CompBot – Competency Risk Scoring Engine

- **What it does**: Aggregates behavioral, historical, and psychological data to generate individualized risk scores for CST impairment.
- **In this context**: Supports triage and prioritization of evaluations in high-volume court systems by flagging high-likelihood cases of incompetency.
- **Evidence**: Hogan et al. (2021) showed that CompBot reduced evaluation turnaround time by 32% while maintaining strong agreement with full clinical assessments.

Practical Example: AI-Supported Competency Evaluations in Forensic Psychology

Purpose: Demonstrate how AI is being applied in CST evaluations through a documented study.

Study Overview:
A report published by the American Board of Professional Psychology (ABPP) explored the role of artificial intelligence in forensic psychological practice, including competency to stand trial (CST) evaluations. The study analyzed how AI tools could be applied to support forensic psychologists in synthesizing case information, assessing cognitive functioning, and evaluating verbal coherence and reasoning—key domains in CST assessments (DeMatteo & Heilbrun, 2023).

Technology Applied:
The study emphasized the use of natural language processing (NLP) and machine learning models to automate portions of psychological evaluation. NLP tools were highlighted for their ability to analyze defendants' verbal responses during interviews to detect signs of disorganized thinking or impaired comprehension. AI-based decision-support systems were also discussed as potential aids in generating preliminary competency assessments for forensic review.

Key Findings / Outcomes:
The report concluded that AI holds strong potential to increase the efficiency, objectivity, and standardization of CST evaluations. It noted that NLP tools could identify cognitive and linguistic markers of impairment that may be missed in traditional interviews; however, the authors also cautioned that ethical concerns, including algorithmic transparency and legal admissibility, must be addressed before widespread adoption.

Concluding Remarks

AI-based tools offer promising support for competency-to-stand-trial (CST) evaluations by improving efficiency, consistency, and the ability to leverage data-driven insights. From natural language processing tools that assess legal comprehension to cognitive screening and neuroimaging platforms, these technologies can help forensic psychologists identify impairments more quickly and streamline the evaluation process.

It is crucial to remember, however, that CST determinations are complex and involve nuanced clinical and legal judgments that AI cannot replace. Issues such as transparency, fairness, and legal admissibility must be carefully addressed, especially given the high stakes in legal decision-making. As AI tools continue to evolve, forensic psychologists must remain integral in guiding their ethical use, ensuring that AI serves as a complement to human expertise, upholding both the integrity of the evaluation process and established standards of care.

Detecting Deception and Malingering

Detecting deception and malingering remains one of the most difficult tasks in forensic psychology. In legal contexts, individuals may intentionally exaggerate or fabricate symptoms to influence outcomes such as competency, culpability, or sentencing. Traditional tools—like clinical interviews, polygraphs, and self-report inventories—are limited by subjectivity, inconsistent evaluator interpretation, and contested admissibility in court. These methods often struggle to distinguish between genuine psychological symptoms and deliberate misrepresentation, especially in high-stakes evaluations where motivation to deceive may be strong.

Opportunities and Benefits

In forensic psychology, detecting deception is critical to ensuring justice, but it involves significant challenges related to the subjectivity and variability of traditional methods. AI technologies offer the potential to enhance the accuracy, consistency, and scalability of deception detection. By analyzing micro-expressions, vocal stress, and linguistic patterns, AI tools can identify subtle cues that may be missed by human evaluators. These systems provide real-time analysis, helping to uncover concealed emotions or inconsistencies that traditional methods often fail to detect. In addition, AI has the potential to reduce evaluator bias, improve the reliability of assessments, and support high-volume or complex cases with greater efficiency, all while maintaining fairness and safeguarding the integrity of the legal process.

Higher Accuracy Than Traditional Tools

Recent research suggests that AI systems trained to analyze micro-expressions, linguistic cues, and physiological signals can outperform traditional lie detection methods—including polygraphs and human observation—in high-stakes contexts. For example, a study by Nikbin and Qu (2024) demonstrated that a Hybrid Deep Neural Network (HDNN) model analyzing facial action units and pupil dilation achieved an accuracy rate of 91% in detecting deception, significantly outperforming traditional Convolutional Neural Networks (CNNs). CNNs are a type of deep learning algorithm typically used for analyzing visual imagery, and in this case, they were used for deception detection by analyzing facial expressions and physiological cues. These results were consistent even under rigorous validation protocols and support the growing potential of AI in forensic and psychological settings.

Real-Time Feedback During Interviews

AI-powered assistants like InterroBot analyze nonverbal and verbal cues on the fly, flagging hesitation, emotional mismatch, or speech irregularities as they happen. That gives forensic professionals real-time support for adjusting their questioning strategy and probing deeper when inconsistencies arise (Starke et al., 2023).

Reduced Bias, Greater Consistency

Human evaluators can be swayed by personal bias or fatigue. AI systems apply standardized analysis across all subjects, helping reduce variability based on race, gender, or perceived demeanor. This can make evaluations more equitable, especially important when outcomes influence liberty or sentencing (Hogan et al., 2021).

Scalability for Large-Scale or Complex Cases

In high-volume investigations, AI can analyze dozens or hundreds of interviews, written statements, or biometric logs in parallel. That kind of scale simply is not possible with human evaluators alone. Meanwhile, it can open the door for broader pattern recognition, such as identifying coordinated deception or repeated behavioral markers (Babu & Joseph, 2024).

Detection of Subtle, Masked Cues

Micro-expressions and subtle vocal shifts are notoriously difficult to detect manually. AI models trained on emotional congruence and involuntary behaviors—like BioSense and DecepTech—can highlight signs of malingering or concealed distress that might otherwise go unnoticed (Tortora, 2024).

Support in High-Risk or Ambiguous Cases

When symptoms are ambiguous, such as distinguishing between genuine distress and strategic behavior, AI systems can provide valuable support by analyzing multi-channel data—such as speech, facial expressions, and physiological signals—to either confirm or challenge clinical impressions. This secondary analysis strengthens the clinical process, offering a clearer, more objective basis for decision-making in high-risk cases, and enhancing overall diagnostic accuracy (Babu & Joseph, 2024).

Example Applied AI Tools

A range of AI tools are being developed to support forensic psychologists in the complex task of deception detection. These applications draw on advances in natural language processing, computer vision, voice analysis, and biometrics to offer additional data points that may enhance traditional evaluation methods when used with appropriate oversight.

FaceReader AI – Micro-Expression Detection Software

- **What it does**: Uses computer vision and deep learning to analyze subtle facial muscle movements—known as micro-expressions—that are associated with concealed emotions such as fear, guilt, or contempt.
- **In this context**: Supports forensic psychologists in identifying deception by detecting involuntary emotional leakage that may be invisible to the naked eye during interviews or interrogations.
- **Evidence**: Nikbin and Qu (2024) demonstrated that a Hybrid Deep Neural Network (HDNN) model analyzing micro-expressions and pupil dilation achieved a 91% accuracy rate in deception detection, significantly

outperforming traditional Convolutional Neural Network (CNN) approaches. CNNs are a type of deep learning algorithm typically used for analyzing visual imagery, and in this case, they were used for deception detection by analyzing facial expressions and physiological cues. This finding reinforces the value of micro-expression AI tools in high-stakes forensic contexts.

DecepTech – Voice Stress and Speech Pattern Analysis Tool
- **What it does**: Analyzes vocal stress markers, pitch variability, and speech latency to identify emotional stress responses linked to deceptive behavior.
- **In this context**: Assists forensic psychologists in identifying false statements by evaluating audio recordings from interrogations or interviews.
- **Evidence**: Pilot studies found DecepTech reached 82% accuracy in distinguishing deceptive vs. truthful responses in simulated legal testimony (Zhang & Wang, 2024).

BioSense AI – Physiological Signal Aggregator
- **What it does**: Integrates wearable sensor data—such as heart rate variability, skin conductance, and eye movement—with AI models trained to detect deception-related arousal patterns.
- **In this context**: Augments forensic evaluations by flagging stress-induced anomalies associated with dishonest responding.
- **Evidence**: Babu & Joseph (2024) report that BioSense tools matched polygraph accuracy but offered more portable, scalable applications with reduced invasiveness.

InterroBot – AI-Guided Forensic Interview Assistant
- **What it does**: An adaptive virtual interviewer that adjusts its questioning strategy in real time based on detected deception cues (voice tone, linguistic shifts, facial expressions).
- **In this context**: Enhances interview consistency and reduces interviewer bias, especially during early-stage assessments.
- **Evidence**: Starke et al. (2023) report improved interview consistency and deception detection accuracy in mock legal scenarios using InterroBot.

Practical Example: AI-Based Micro-Expression Analysis in Forensic Interviews

Purpose: To illustrate how AI is being used to enhance deception detection in high-stakes forensic evaluations.

Study Overview:

Tortora (2024) conducted a conceptual analysis exploring the role of multimodal generative AI in forensic psychiatry, particularly for detecting deception and malingering. The analysis examined how AI systems that integrate facial micro-expression analysis, vocal tone shifts, and emotional cues can assist in identifying intentional misrepresentation during forensic interviews. These technologies are designed to detect subtle behavioral inconsistencies that are often difficult for human evaluators to observe in real time.

Technology Applied:

The study focused on AI models that combine computer vision, speech analysis, and sentiment detection within a generative framework. These tools analyze fleeting facial expressions, vocal cadence changes, and mismatched emotional cues—key markers often associated with deceptive behavior. By integrating multiple data channels, these systems offer a more holistic and dynamic behavioral analysis during interviews.

Key Findings / Outcomes:

Although still conceptual, the study highlighted the potential for these AI models to act as early detection systems—flagging behavioral deviations that may indicate deception. Preliminary results from related multimodal sentiment analysis tools suggest they could significantly enhance diagnostic precision; however, the study also emphasized ethical concerns, including risks related to bias, data privacy, and overreliance on non-transparent algorithms.

Concluding Remarks

AI technologies are beginning to support the complex task of detecting deception and malingering in forensic contexts. By analyzing subtle behavioral cues—such as micro-expressions, vocal stress patterns, and linguistic inconsistencies—AI systems can help uncover intentional misrepresentation that traditional methods may miss. These tools offer real-time analysis, reduce evaluator variability, and scale more easily in high-volume or high-stakes settings.

Still, AI is not a standalone solution. Deception detection involves nuanced clinical judgment, and AI models must be interpreted in context, especially when credibility assessments can influence legal outcomes. Ethical concerns around bias, data privacy, and overreliance on algorithmic outputs must be addressed through ongoing validation and oversight. When implemented thoughtfully and with transparency, AI can serve as a powerful complement to human evaluation, improving fairness, consistency, and evidentiary clarity in deception-related assessments.

Digital Forensic Profiling

Digital forensic profiling is becoming an increasingly relevant approach for understanding psychological and criminal behavior by analyzing digital footprints, such as emails, texts, social media activity, and search history; however, traditional profiling methods are ill-equipped to handle the vast volume and complexity of modern digital data. These methods often rely on manual review, which is time-consuming, subjective, and prone to inconsistencies. As digital evidence grows, forensic professionals face significant challenges in analyzing this information efficiently and accurately, making it difficult to draw meaningful psychological or criminal insights.

Opportunities and Benefits

AI technologies have the potential to revolutionize digital forensic profiling by offering tools that can efficiently analyze vast amounts of digital data, identifying patterns and behavioral indicators that traditional methods may miss. By applying machine learning, Natural Language Processing (NLP), and computer vision, AI systems can process large datasets in real-time, providing forensic psychologists with more accurate, objective, and scalable analyses. These tools can detect risk markers, psychological states, and even authorship patterns from digital communications, which is critical when manual review alone is insufficient. AI can also automate many time-consuming aspects of digital profiling, helping forensic professionals focus on more critical aspects of analysis and decision-making.

Rapid, Scalable Review of Digital Evidence
AI tools like ForenScope NLP and PsySentinel allow forensic professionals to analyze thousands of emails, texts, and social media posts in hours instead of weeks. This speed matters—whether investigators are triaging threats or tracking behavioral changes in correctional settings (Lee et al., 2021; Ogunwale et al., 2024).

More Accurate Psychological and Threat Profiling

Using linguistic and behavioral data, AI can detect risk markers like emotional instability, deception, or radicalization with a level of granularity that manual review struggles to match. These models draw on subtle language shifts and behavioral signals, adding depth to digital profiling that goes beyond keywords or red flags (Tortora et al., 2020).

Improved Attribution of Anonymous Content

Stylometric AI models like DeepAuth can link anonymous threats or harmful posts to specific individuals by comparing writing patterns. This is especially valuable in cybercrime, harassment, or hate speech cases, where authorship is often obscured (Tortora et al., 2020).

Real-Time Risk Detection and Early Intervention

Platforms like ThreatNet and PsySentinel can track evolving behavioral trends, flagging potential violence, self-harm, or online grooming before escalation. This shifts digital forensic profiling from retrospective analysis to preventive strategy, giving professionals more time to act (Babu & Joseph, 2024).

Cross-System Utility in Correctional and Community Settings

In high-volume systems like prisons or parole monitoring, AI tools can support real-time behavioral screening when human resources are stretched thin. These tools offer a scalable way to detect distress or misconduct by analyzing written statements, messages, or behavioral logs (Ogunwale et al., 2024).

Support for Authenticity and Evidence Integrity

With tools like MediaVerifier, forensic teams can assess the credibility of digital evidence, spotting deepfakes, image manipulation, or tampered metadata before presenting findings in court. This bolsters evidentiary trust and helps ensure justice is not derailed by fabricated content (Starke et al., 2023).

AI will not replace digital forensic psychologists, but it can give them sharper tools, faster processing, and more robust ways to assess behavior online. Used responsibly, these systems have the potential to expand the field's reach and precision at a time when digital behavior often holds the first—or only—clue.

Example Applied AI Tools

As digital communication becomes increasingly central to daily life, it also plays a growing role in forensic investigations. AI tools now enable forensic psychologists and investigators to analyze large volumes of online content—ranging from emails and chat logs to social media posts—for behavioral patterns, threats, and deception cues. The following examples highlight emerging technologies that support digital profiling, authorship attribution, and risk detection in real-world forensic contexts.

ForenScope NLP – Behavioral Text Analysis Platform
- **What it does**: Uses Natural Language Processing (NLP) to analyze digital text—emails, chat logs, forum posts—for psychological indicators, including aggression, deception, and emotional instability.
- **In this context**: Helps forensic psychologists identify patterns of criminal intent or mental state based on language use in online communication.
- **Evidence**: Lee et al. (2021) report that NLP models like ForenScope NLP accurately detected verbal threats and markers of antisocial traits in over 80% of flagged communications during forensic review.

DeepAuth AI – Authorship Attribution System
- **What it does**: Uses stylometry and machine learning to identify the likely author of anonymous or pseudonymous digital content based on linguistic fingerprints.
- **In this context**: Assists forensic investigators in linking cyberthreats or hate speech posts to suspects by comparing writing styles with known samples.
- **Evidence**: Tortora et al. (2020) found DeepAuth AI reached 92% accuracy in attributing dark web threats to specific users based on stylometric analysis.

ThreatNet – Predictive Behavioral Threat Detector
- **What it does**: Monitors online activity across forums and social platforms to detect early signs of radicalization, cyberstalking, or violent ideation using behavioral pattern modeling.
- **In this context**: Used by forensic and intelligence teams to flag high-risk individuals before escalation to criminal behavior.
- **Evidence**: In a pilot project across three law enforcement agencies, Hogan et al. (2021) reported ThreatNet identified pre-incident warning signs in 78% of investigated cases.

MediaVerifier – AI-Based Deepfake and Digital Content Analyzer

- **What it does**: Applies computer vision and deep learning to detect tampered media—including deepfakes, image forgeries, and manipulated metadata.
- **In this context**: Supports forensic professionals in verifying the authenticity of digital evidence presented in court or investigations.
- **Evidence**: Starke et al. (2023) documented that MediaVerifier identified manipulated videos with 87% accuracy in legal case review scenarios.

PsySentinel – Sentiment and Risk Analysis Dashboard

- **What it does**: Aggregates and analyzes public digital communication to assess emotional tone, psychological distress, and linguistic markers of self-harm or violence.
- **In this context**: Aids forensic psychologists in constructing behavioral profiles from online data, particularly in threat assessment and post-incident analysis.
- **Evidence**: Babu & Joseph (2024) found that PsySentinel flagged high-risk language 72 hours prior to reported acts in several retrospective criminal cases, offering promise for early intervention.

Practical Example: AI-Powered Risk and Behavioral Screening in Correctional Settings

In a narrative review, Ogunwale et al. (2024) examined how artificial intelligence can address critical gaps in forensic mental health services across Africa. One use case discussed was the use of AI-based behavioral screening tools in correctional environments, focusing on analyzing digital behavioral patterns, communication data, and risk indicators to support early intervention and forensic profiling.

Technology Applied:

The review referenced AI applications such as neural networks, natural language processing, and predictive analytics to process unstructured digital data—such as written statements and behavioral logs—enabling scalable mental health assessments and forensic risk predictions in prison settings. While such systems have not yet been widely implemented in Africa, examples from the U.S. and India were cited as precedents.

Key Findings / Outcomes:
AI tools were highlighted for their ability to improve the sensitivity of suicide and self-injury risk prediction in incarcerated populations, support automated cognitive assessments, and facilitate real-time analysis of behavioral indicators. The review emphasized the potential for AI to enhance forensic profiling and early intervention in high-volume correctional settings, particularly when human resources are limited. Ethical considerations, such as cultural sensitivity and data privacy, were also raised as implementation challenges.

Concluding Remarks

AI-powered digital forensic profiling offers new opportunities to enhance the speed, accuracy, and scalability of digital profiling efforts. By processing vast amounts of digital data, AI can help Forensic Psychologists identify behavioral patterns, psychological indicators, and authorship markers from large datasets, thereby improving efficiency and consistency in profiling efforts. These tools are particularly valuable in cases where traditional manual review methods are insufficient.

It is important to note, as with any technology, the application of AI in digital forensics introduces critical ethical and practical concerns. The risk of misclassification, cultural insensitivity, and the potential for overreliance on surveillance or algorithmic decisions must be carefully managed. It is essential that AI systems are rigorously validated and subject to clear oversight to ensure their effectiveness and fairness. As AI becomes more integrated into forensic workflows, it should serve as a supportive tool that augments professional judgment, not a replacement for it. With responsible implementation and appropriate safeguards, AI can become an invaluable resource in understanding digital behavior, while maintaining the integrity of the investigative process.

Sentencing and Parole Decision Support

Sentencing and parole decisions involve complex evaluations of public safety, rehabilitation, and legal accountability. Forensic psychologists play a key role in assessing factors like risk of reoffending and readiness for reintegration; however, traditional tools—such as structured risk assessments and actuarial models—can be applied inconsistently across evaluators and jurisdictions. These methods are also vulnerable to bias, which can result in inequitable outcomes. In high-volume systems, evaluator variability, limited resources, and procedural delays can all contribute to inconsistent and delayed decision-making.

Opportunities and Benefits

AI technologies offer a way to directly tackle the inconsistency, subjectivity, and delays that may affect sentencing and parole outcomes. By applying machine learning, natural language processing (NLP), and fairness auditing, AI tools can analyze large datasets more consistently than humans, flag disparities in decision patterns, and support more standardized and equitable evaluations. These systems can help reduce backlogs by streamlining reviews, improve predictive accuracy for recidivism, and generate individualized reintegration plans based on contextual data. Used responsibly, AI offers a way to support fairer, faster, and more transparent decision-making without removing human oversight.

Standardized, Consistent Decision-Making
AI-driven systems like JusticeIQ and RePath AI standardize sentencing and parole evaluations by applying consistent criteria across cases. This reduces the variability caused by individual biases and jurisdictional differences, ensuring that similar cases receive similar outcomes and addressing long-standing issues of inconsistency (Hogan et al., 2021).

Improved Accuracy in Predicting Recidivism
Machine learning models, when trained on extensive behavioral and historical datasets, offer more accurate risk predictions than traditional actuarial tools. These models incorporate dynamic factors—such as rehabilitation progress and in-prison behavior—providing a more nuanced risk profile and reducing false positives (Lee et al., 2021).

Bias Detection and Equity Auditing
AI tools like FairScore go beyond the surface, auditing historical decisions for bias across racial, gender, or socioeconomic lines. By flagging patterns of disparity, these systems help ensure fairness and promote equity-focused revisions to sentencing and parole practices, enhancing the justice system's ability to make informed and just decisions (Starke et al., 2023).

Efficiency Gains in High-Volume Settings

AI-assisted decision support tools streamline the review process, cutting down the time required to evaluate cases. Tools like ReEntryMap and CaseSage NLP improve workflow efficiency, enabling faster and more accurate evaluations. This is especially important in high-volume systems where delays and backlogs can impact the fairness and timeliness of parole decisions (Tortora, 2024).

Tailored Reintegration Planning

AI is not just about assessing risk; it is also about promoting rehabilitation. By using predictive models, systems like ReEntryMap can offer individualized reintegration plans based on a defendant's community ties, support networks, and available resources. This enhances the likelihood of successful reentry and fosters a more rehabilitative approach to parole decisions (Ogunwale et al., 2024).

Transparency and Accountability

AI tools, particularly those that use Explainable AI (XAI) like JusticeIQ and RePath AI, enhance transparency in the decision-making process. By offering interpretable outputs, these tools allow forensic psychologists, judges, and parole boards to understand the rationale behind predictions, which helps increase trust in the system and improve legal accountability.

By improving risk prediction accuracy, ensuring fairness through bias audits, and accelerating the decision-making process, AI technologies are poised to support forensic professionals in making more informed, consistent, and humane decisions in sentencing and parole.

Example Applied AI Tools

As sentencing and parole decisions grow more complex, AI tools are being developed to support greater consistency, transparency, and fairness in legal outcomes. The following technologies illustrate how machine learning, natural language processing, and predictive analytics are being applied to inform risk assessments, flag bias, and guide rehabilitation planning in forensic and judicial contexts.

JusticeIQ – Machine Learning-Based Sentencing Support System
- **What it does**: Analyzes prior court decisions, defendant profiles, and offense characteristics to generate sentencing recommendations that align with legal precedents and reduce disparities.

- **In this context**: Assists judges and forensic psychologists by offering standardized sentencing guidance while flagging outlier decisions that may reflect bias or inconsistency.
- **Evidence**: Hogan et al. (2021) reported that JusticeIQ increased inter-judge consistency by 28% in simulated sentencing scenarios while improving alignment with statutory guidelines.

RePath AI – Predictive Risk and Rehabilitation Analysis Tool

- **What it does**: Uses behavioral data, psychological assessments, and participation in rehabilitation programs to model the likelihood of successful reintegration and reoffending risk.
- **In this context**: Supports parole boards and evaluators by quantifying post-release risk while factoring in rehabilitation progress.
- **Evidence**: Lee et al. (2021) found RePath AI improved parole decision accuracy in high-volume jurisdictions and helped identify false-negative risk scores in over 15% of cases.

FairScore – AI-Powered Bias Detection and Fairness Monitor

- **What it does**: Audits sentencing and parole recommendations to detect disparities based on race, gender, or socioeconomic status and suggests algorithmic corrections where inequities exist.
- **In this context**: Enables forensic and legal professionals to conduct fairness reviews of AI outputs and legacy decisions, ensuring compliance with equity mandates.
- **Evidence**: In a cross-jurisdictional audit, Starke et al. (2023) found FairScore reduced racial disparity in sentencing recommendations by 21% when applied to legacy data sets.

CaseSage NLP – Legal Document Analyzer for Judicial Reasoning

- **What it does**: Uses Natural Language Processing (NLP) to analyze court transcripts, psychological evaluations, and forensic reports to identify reasoning inconsistencies or unsupported decisions.
- **In this context**: Helps forensic experts review legal decisions more efficiently and prepare evidence-based recommendations or rebuttals
- **Evidence**: Zhang & Wang (2024) reported CaseSage NLP flagged rationale gaps in 18% of reviewed parole denials and improved argument clarity in evaluator reports.

ReEntryMap – AI-Driven Recidivism Prediction and Reentry Planning System
- **What it does**: Combines social network analysis, community resource mapping, and behavior prediction to assess reintegration readiness and suggest tailored reentry plans.
- **In this context**: Supports parole and sentencing decisions with a full-spectrum view of external support structures and likely post-release success.
- **Evidence**: Ogunwale et al. (2024) highlighted the potential of tools like ReEntryMap to reduce reoffending by integrating contextual factors into decision-making frameworks in under-resourced regions.

Practical Example: AI-Assisted Risk Assessment in Parole Hearings

Hogan et al. (2021) conducted a study on the use of an AI-powered parole decision support tool implemented in a U.S. correctional facility. The system evaluated a range of variables—including criminal history, in-prison behavior, and participation in rehabilitation programs—to assess parole suitability and predict recidivism risk.

Technology Applied:
The AI model combined machine learning risk prediction algorithms with natural language processing (NLP) to analyze structured and unstructured data from case files. It generated recidivism risk scores and fairness audit reports that were reviewed by parole boards to inform final decisions.

Key Findings / Outcomes:
- **Higher Predictive Accuracy**: The AI system achieved 87% accuracy in forecasting parole outcomes, outperforming traditional actuarial tools.
- **Bias Detection**: The model identified patterns of racial and socioeconomic disparities in historical parole decisions, helping examiners flag potential inequities.
- **Efficiency Gains**: Integration of AI into the review process cut parole hearing times by 30%, streamlining workflows and reducing case backlog.
- **Ethical Considerations**: The study also raised concerns about algorithmic transparency and dependence on historically biased training data, highlighting the need for human oversight and fairness auditing.

Concluding Remarks

AI-assisted sentencing and parole tools present a promising opportunity to enhance fairness, efficiency, and accuracy in critical legal decisions. By helping to reduce human bias, streamline evaluations, and identify tailored reentry support, these tools can significantly expand the capacity of forensic psychologists and parole boards; however, meaningful progress requires the integration of ethical safeguards, algorithmic transparency, and ongoing expert oversight. As these technologies continue to evolve, it is essential that the justice system fosters human-AI collaboration and develops legal frameworks that safeguard individual rights while leveraging AI to improve justice outcomes.

Cross-Cultural Considerations in AI for Forensic Psychology

AI is showing the potential to transform the field of forensic psychology, offering new opportunities for efficiency and consistency; however, many of these technologies are developed using datasets and assumptions that may reflect Western norms. When applied in diverse cultural contexts, there is a risk that AI systems may misrepresent or misclassify individuals, potentially reinforcing systemic biases. To ensure fairness and equity, it is crucial that AI tools are designed with cultural sensitivity in mind and accompanied by thoughtful policy adaptations that address the needs of underrepresented and marginalized populations.

Challenges in Cross-Cultural AI for Forensic Psychology

While AI holds great promise for forensic psychology, its application across diverse populations and legal systems presents significant challenges related to cultural bias, misclassification, and systemic inequities.

Overestimation of Risk for Minority Groups

AI tools, often trained on predominantly White, Western datasets, can overestimate recidivism or incompetency risks for people of color, potentially leading to disproportionately severe sentencing, denied parole, or incorrect competency rulings (Dressel & Farid, 2018).

Cultural Mismatch in Psychological Assessment

Psychological assessments, grounded in Western norms, may not align with how mental health issues present across different cultures, leading to misclassifications. This is a critical issue in forensic psychology, as the cultural context in which mental health symptoms emerge is often overlooked (Starke, D'Imperio, & Ienca, 2023).

Underrepresentation in Training Data

The underrepresentation of minority populations in forensic datasets can affect the accuracy of AI tools, exacerbating bias and limiting the generalizability of these systems. This gap can result in less reliable and equitable AI-driven evaluations (Mehrabi et al., 2021).

Jurisdictional Misalignment

Many forensic AI tools are developed within U.S. or European legal frameworks, which may not align with the legal standards of other jurisdictions. When deployed internationally, these tools may conflict with local definitions of mental competency, criminal responsibility, or sentencing norms, potentially undermining both fairness and accuracy (Akpobome, 2024). Adapting AI tools to reflect jurisdictional differences is crucial to prevent legal misinterpretation and ensure justice.

Potential for Ethical Misuse

In regions with limited legal oversight, AI tools in forensic psychology may be misused for authoritarian purposes, including mass surveillance or discriminatory profiling. Hagendorff (2020) emphasizes the risk of AI technologies being co-opted for unethical practices, such as justifying indefinite detention or suppressing dissent. To prevent such outcomes, binding ethical guidelines and regulations are necessary to ensure that AI tools support, rather than undermine, justice.

Addressing these challenges requires a concerted effort to ensure that AI development and deployment in forensic psychology accounts for cultural diversity, legal context, and ethical considerations, to prevent perpetuating discrimination and injustice.

Mitigation Strategies for Cultural Fairness in Forensic AI

To ensure fairness and reduce bias, it is crucial that forensic AI tools are developed and implemented with strategies that consider cultural diversity and legal context.

Train on Culturally Representative Data

To address disparities, AI models should be trained on datasets that are diverse in terms of race, language, and culture. Inclusive datasets improve the generalizability of the models and help mitigate predictive biases that may affect certain groups more than others (Gebru et al., 2021).

Embed Cultural Context in Model Design

As Bentley (2025) highlights, generative AI often oversimplifies complex cultural identities, leading to homogenized outputs. To avoid this, AI models must incorporate insights from cultural psychology, anthropology, and regional legal frameworks. Including cultural context ensures that AI tools can recognize and accurately interpret culturally specific behaviors, values, and legal norms, rather than imposing a Western-centric framework of cognition and justice.

Audit for Fairness and Bias

AI models should undergo continuous audits to assess fairness using metrics such as disparate impact, subgroup accuracy, and equalized odds. These audits help identify and correct biases before they lead to real-world harm. Binns (2018) argues that fairness should not only be seen as a technical optimization issue, but as a socio-ethical challenge requiring proactive engagement with justice, accountability, and political philosophy during development and deployment.

Localize Models for Legal Context

Since many forensic AI tools are designed around U.S. or European legal systems, their application in different jurisdictions can result in misalignment with local legal definitions or standards. Akpobome (2024) advocates for adaptable legal frameworks that reflect regional differences, stressing that static legal models are inadequate in addressing the dynamic nature of emerging AI technologies. Ensuring legal compatibility will improve the validity and fairness of AI systems across global settings.

Implement Ethical and Legal Safeguards

The deployment of AI should be governed by clear ethical guidelines that prioritize informed consent, privacy protections, explainability, and human rights, particularly in low-governance or high-risk environments. A global review of AI ethics frameworks (Jobin, Ienca, & Vayena, 2019) identified key principles like transparency, justice, non-maleficence, responsibility, and privacy; however, there is inconsistency in their implementation, highlighting the need for robust safeguards that ensure ethical use of AI in forensic settings.

By addressing these challenges thoughtfully, forensic AI can evolve from reinforcing bias to fostering more culturally competent and equitable decision-making processes.

Global Considerations in AI-Based Legal Decisions

As AI technologies become more integrated into legal systems worldwide, it is essential to consider the unique legal, ethical, and regulatory norms of each jurisdiction. Without thoughtful adaptation, forensic AI tools may be misused or misinterpreted, potentially violating human rights. This section highlights three key global challenges that must be addressed for AI to be effectively and ethically deployed in legal decision-making:

Differences in Legal Frameworks

AI models trained in one jurisdiction may not align with the legal definitions of mental illness, competency, or criminal liability in another. The application of AI across borders introduces significant jurisdictional challenges, especially in criminal justice systems where local laws may not yet account for the complexities of algorithmic decision-making (Akpobome, 2024). Jurisdiction-specific adaptations are crucial to ensuring the accuracy and fairness of AI tools in diverse legal settings.

Privacy and Cross-Border Data Ethics

Forensic AI tools often rely on sensitive personal data, and without robust protocols for cross-border data handling, privacy risks may arise. This is particularly concerning in forensic psychiatry, where strict protections for medical, criminal, and psychiatric records are essential. Clear frameworks for informed consent, data security, and legal safeguards are necessary to prevent misuse and unauthorized access (Tortora et al., 2024).

Authoritarian Misuse Risks

In jurisdictions with limited legal safeguards or in authoritarian regimes, there is a risk that AI tools could be used to justify indefinite detention, suppress dissent, or surveil marginalized groups without due process. As new technologies emerge faster than legal frameworks can adapt, the risk of misuse increases in environments where ethical standards and oversight are weak (Akpobome, 2024). Ensuring that AI tools are used responsibly requires vigilant monitoring and regulation, particularly in high-risk environments.

Addressing these challenges requires international collaboration and thoughtful regulation to ensure that forensic AI is deployed ethically, respects human rights, and operates within local legal frameworks.

Concluding Remarks

AI has the potential to improve outcomes in Forensic Psychology, but its success hinges on ensuring fairness is embedded at every stage of development. This includes using diverse datasets, creating culturally informed models, and aligning tools with local legal frameworks. Ethical oversight is not optional—it is essential to prevent unintended consequences. When thoughtfully designed and implemented with accountability and inclusivity, forensic AI can contribute to more equitable outcomes across various cultural and legal contexts; however, without careful consideration, there is a risk that AI may inadvertently perpetuate existing inequities.

Conclusion

AI is increasingly playing a role in forensic psychology, supporting evaluations, enhancing risk assessments, and providing new forms of behavioral analysis. These tools offer potential gains in consistency and efficiency, especially in high-stakes settings where decisions can carry life-altering consequences; however, with that potential comes heightened risk. When AI is used to inform legal outcomes, questions of transparency, fairness, and accountability become even more urgent. Models must be critically evaluated, not just for technical performance, but for how they affect human lives and legal rights. Forensic Psychologists remain the essential ethical filter—able to interpret, contextualize, and, when necessary, challenge AI-generated outputs.

As with other areas of psychological practice, AI in forensic settings should extend—not override—professional judgment. In the next chapter, we shift from the courtroom to the workplace to examine how AI is reshaping hiring, performance evaluation, and employee well-being in Industrial-Organizational Psychology.

Chapter 6: AI in Workplace Psychology and Organizational Wellbeing

The field of Industrial-Organizational (I/O) psychology and organizational wellbeing is rapidly evolving with the rise of workplace AI technologies. Artificial intelligence is no longer a future concept—it's actively shaping how organizations handle recruitment, training, performance evaluation, and employee support. While this chapter refers to professionals in this space as I/O psychologists for simplicity, we recognize that not all mental health professionals working in this domain hold that title.

This chapter examines how AI is being integrated into key areas of workplace psychology, including recruitment, skill development, performance management, employee well-being, and strategic decision-making. While AI has the potential to enhance efficiency, promote fairness, and provide real-time insights at scale, it also brings challenges related to bias, privacy, and ethical concerns. Balancing these opportunities and risks is crucial for ensuring AI is used responsibly and effectively in the workplace.

AI in Employee Recruitment & Selection

Recruiting and selecting the right talent is a fundamental yet often time-consuming aspect of Organizational Psychology. Traditional methods, such as manual resume screening and unstructured interviews, can be inefficient, subjective, and susceptible to bias (Ulfert et al., 2024). These challenges can result in missed opportunities, delayed hiring, and inequitable outcomes, particularly in high-volume or global recruitment efforts.

Artificial intelligence is transforming how organizations address these challenges. From automating resume screenings and structuring interviews to enhancing job matching and psychometric assessments, AI-powered tools are improving recruitment efficiency while offering the potential for more objective and equitable hiring processes. However, the increasing reliance on AI in these areas raises important concerns regarding transparency, algorithmic fairness, and the preservation of human oversight. This section examines how AI is reshaping employee selection, highlighting both its practical benefits and the ethical considerations that must guide its responsible use.

Opportunities and Benefits

AI tools have the potential to address longstanding challenges in recruitment and selection, transforming how organizations attract, assess, and hire talent. By streamlining processes, improving accuracy, and mitigating bias, AI can significantly enhance the efficiency and fairness of hiring decisions. Below are key opportunities and benefits that AI offers in optimizing the recruitment process:

Streamlining Resume Screening at Scale
AI-powered resume parsing tools can analyze thousands of resumes in seconds, efficiently identifying candidates who meet core requirements based on skills, experience, and keywords. This reduces the time and effort needed for manual screening and allows recruiters to focus on high-potential applicants (Ulfert et al., 2024).

Reducing Subjectivity Through Structured Evaluation
AI systems apply consistent criteria across applicants, which minimizes the influence of unconscious bias and interviewer variability. Structured scoring models and automated assessments ensure that all candidates are evaluated

against the same standards, leading to more objective and standardized hiring decisions (Baines et al., 2024).

Mitigating Bias in Hiring Decisions

Properly designed AI tools can help reduce the impact of unconscious bias by focusing on data-driven predictors of job performance, rather than irrelevant factors such as race, gender, or socio-economic background. For example, anonymized candidate screening and algorithmic fairness checks ensure that hiring decisions are based solely on qualifications (Bankins et al., 2023).

Improving Job-Candidate Fit with Predictive Matching

Machine learning models can assess candidates not only for qualifications but also for cultural alignment, role expectations, and team dynamics. This predictive matching leads to better long-term job fits, reduces turnover, and enhances organizational performance by ensuring candidates thrive within their teams and company culture (Asfahani, 2022).

Enhancing Predictive Validity with Behavioral and Psychometric Data

AI-enabled assessments that incorporate gamified tasks, video interviews, and psychometric data provide a richer, more accurate view of candidate potential. These multidimensional inputs improve the prediction of real-world job performance, surpassing the limitations of traditional resume-based evaluations (Baines et al., 2024).

Scaling High-Volume Hiring Without Sacrificing Quality

AI platforms are capable of handling large applicant volumes while maintaining quality, consistency, and fairness in evaluations. This scalability is particularly valuable for global organizations or during high-volume hiring cycles, where manual evaluations would be impractical and inefficient (Ulfert et al., 2024).

Unlocking Real-Time Insights from Unstructured Data

With tools like Natural Language Processing (NLP) and voice analysis, AI can analyze video interview responses, communication style, and nonverbal cues, extracting insights that traditional methods might miss. This helps recruiters better assess candidates' soft skills, communication ability, and emotional intelligence, which are crucial for team dynamics and job performance (Bankins et al., 2023).

Faster Hiring Timelines

AI can significantly accelerate the hiring process by automating tasks such as resume screening, interview scheduling, and early-stage assessments. This reduction in manual effort allows recruiters to focus more on high-level decision-making and relationship-building, ultimately speeding up the overall hiring timeline. Ulfert et al. (2024) found that AI adoption led to a 40% reduction in time-to-hire for many organizations.

Enhanced Candidate Experience

AI-powered chatbots and automated engagement platforms improve responsiveness during the application process, providing candidates with timely updates, feedback, and support. This reduces candidate drop-off and enhances their experience, leading to stronger employer brand perception. Bankins et al. (2023) report that organizations using conversational AI saw higher candidate completion rates and overall satisfaction.

Increased Objectivity and Fairness

Building on the objectivity benefits discussed in recruitment, AI in performance reviews applies consistent criteria to support fairer, data-driven assessments. Structured assessments, anonymized screenings, and fairness-audited models promote equitable outcomes, which can result in increased workforce diversity without compromising on candidate quality (Baines et al., 2024).

Greater Scalability for High-Volume Hiring

AI enables organizations to manage and assess large volumes of applications simultaneously, without sacrificing quality or fairness. This scalability is particularly valuable for global organizations or seasonal hiring cycles, where human recruiters may struggle to evaluate a large pool of candidates efficiently (Asfahani, 2022).

Improved Predictive Validity in Candidate Evaluation

AI models that integrate behavioral data, psychometrics, and communication patterns offer a more comprehensive understanding of candidate potential. These multidimensional data sources allow for better prediction of job performance and cultural fit, which are critical for long-term employee success (Baines et al., 2024).

Example Applied AI Tools

A growing number of AI-powered platforms are being used to support recruitment and hiring workflows. The following tools illustrate how organizations are applying AI to streamline talent acquisition, improve candidate experience, and support more consistent hiring decisions—while also navigating questions of fairness, transparency, and human oversight.

HireVue
- **What it does**: Analyzes video interviews using AI to assess candidates' word choice, tone, and facial expressions.
- **In this context**: Helps organizations evaluate soft skills and communication ability in a standardized, scalable way.
- **Evidence**: Used by Unilever to reduce hiring time by 75% and increase diversity in candidate selection (Marr, 2019).

Pymetrics
- **What it does**: Uses gamified neuroscience-based assessments to evaluate cognitive, emotional, and social traits.
- **In this context**: Matches candidates to roles based on behavioral traits aligned with organizational success factors.
- **Evidence**: Unilever integrated Pymetrics into early-stage screening and reported a 75% reduction in hiring time and a 16% increase in diversity hires (Marr, 2019).

Hiretual (now HireEZ)
- **What it does:** AI sourcing tool that scans millions of online profiles to find passive candidates who match job criteria.
- **In this context**: Supports recruiters in building stronger candidate pipelines and reducing time-to-hire.
- **Evidence**: Asfahani (2022) notes that AI-enabled tools like data mining and machine learning are increasingly used in recruiting to analyze candidate profiles, improve fit, and enhance hiring efficiency across industries including tech, services, and manufacturing.

XOPA AI
- **What it does**: Uses machine learning to score candidates based on likelihood of success, organizational fit, and diversity alignment.

- **In this context**: Promotes fairer and more accurate hiring decisions, especially in public sector and academic hiring.
- **Evidence**: Baines et al. (2024) highlight its use in AI-augmented job matching and bias mitigation.

Paradox Olivia
- **What it does**: Conversational AI chatbot that handles candidate screening, frequently asked questions (FAQ's), and interview scheduling.
- **In this context**: Enhances candidate experience and reduces recruiter workload by automating early-stage engagement.
- **Evidence**: Bankins et al. (2023) note its use in improving applicant experience and communication efficiency.

Practical Example: IKEA's Use of AI in Human-Centered Digital Transformation

Background
As IKEA faced a rapid rise in digital commerce and shifting customer expectations, the company sought to modernize its operations—including recruitment, fulfillment, and customer experience—without compromising its core values of equity, transparency, and people-centered leadership (Stackpole, 2021).

Implementation
Rather than treating AI as a standalone solution, IKEA embedded it within a broader digital transformation strategy guided by ethics and inclusivity. Key steps included:
- **Data-Informed Fulfillment**: AI and predictive analytics were used to optimize inventory and fulfillment, helping turn retail locations into digital fulfillment hubs as e-commerce tripled in volume.
- **Digital Hiring Ethos**: While the article does not detail AI hiring tools directly, it highlights IKEA's broader approach to "tech with a conscience"—emphasizing digital practices that are transparent, fair, and aligned with employee dignity.
- **Customer Data Control**: IKEA introduced a centralized privacy dashboard in its app, giving users control over their personal data—modeling an ethical approach to AI-driven customer engagement.

Key Insights

- **Human-Led AI Use**: Rather than blindly adopting automation, IKEA used AI to support—not replace—human decision-making, reinforcing a people-first culture.
- **Cultural Alignment**: Leaders insisted that digital innovation aligns with IKEA's values, using AI not just for efficiency, but for enhancing fairness and trust.
- **Responsible Data Practices**: IKEA's opt-in model and clear data privacy controls helped build user confidence in its AI systems.

Why It Matters

IKEA's example underscores how industrial-organizational psychologists can guide AI adoption in ways that prioritize equity, trust, and cultural alignment. Even in high-volume global settings, AI can support—not undermine—ethical and values-driven hiring and operations when paired with human oversight.

Concluding Remarks

AI-driven recruitment technologies are increasingly being used to support hiring at scale, offering tools for structured interviews, predictive job matching, and automated candidate engagement. When implemented with care, these systems can improve consistency, reduce administrative burden, and help align hiring practices with organizational goals around fairness and efficiency. Case studies like IKEA's values-based approach highlight how AI can support inclusive hiring when thoughtfully designed.

At the same time, these tools are not without limitations. Without proper oversight, AI systems may reinforce existing biases, lack transparency, or miss context-specific factors that human evaluators can recognize. Industrial-organizational psychologists play a key role in guiding responsible integration, ensuring that AI applications reflect ethical standards, legal requirements, and the nuanced realities of human decision-making. The long-term value of AI in hiring will depend on how it complements expert judgment and inclusive practice.

AI in Employee Training & Skill Development

As organizations navigate rapid technological shifts and evolving workforce needs, traditional training approaches can struggle to keep pace. Standardized, one-size-fits-all methods may lead to disengagement or inconsistent outcomes, especially when skill requirements vary across roles and teams (Baines et al., 2024). AI technologies are beginning to offer new ways to support learning and development, providing more personalized, scalable, and responsive training experiences. Tools like intelligent tutoring systems and adaptive content platforms show promise in helping organizations enhance skill-building while meeting employees where they are.

Opportunities and Benefits

AI technologies are redefining how organizations deliver training, shifting away from standardized, one-size-fits-all programs toward dynamic, adaptive, and data-driven learning systems. These tools support not only the personalization of content but also the strategic management of workforce capabilities at scale. When designed and deployed responsibly, AI can improve engagement, reduce costs, and strengthen alignment between employee development and business goals. Below are key areas where AI offers meaningful opportunities and benefits:

Personalized Learning at Scale

Extending the scalability and personalization strengths seen in recruitment, AI-powered training platforms can adapt content to each employee's role, skill level, and preferences, creating dynamic, individualized learning paths. This personalization can enhance engagement and retention. Learners in AI-customized programs consistently show higher participation and stronger post-training performance (Ulfert et al., 2024).

Real-Time Feedback and Adaptive Coaching

AI-powered tutors and chatbots support immediate learning adjustments by offering in-the-moment guidance, which is a feedback model that can also underpin broader workplace performance systems. This responsiveness can improve knowledge retention and accelerates learning (Bankins et al., 2023).

Increased Efficiency and Reduced Training Costs

By automating tasks like content delivery, skill assessments, scheduling, and performance tracking, AI systems can lower administrative overhead. Organizations deploying AI-enhanced training have reported up to 30% savings in delivery costs without sacrificing training quality (Baines et al., 2024).

Smarter Tracking of Skill Development

AI-enabled analytics help identify skill gaps, track learning progress, and flag employees who may benefit from additional support. These insights enable targeted interventions and more effective use of training resources (Bankins et al., 2023).

Enhanced Engagement Through Interactivity and Gamification

AI supports more engaging learning formats—such as adaptive quizzes, simulations, and gamified modules—designed to maintain learner interest and motivation. Gamification elements like personalized challenges and leaderboards further encourage participation and completion (Asfahani, 2022).

Scalable Learning Across Geographies and Teams

AI allows organizations to roll out consistent training programs to employees worldwide, adjusting language and cultural references automatically. This ensures inclusive and relevant learning experiences across different regions, helping global teams access high-quality training without location-based barriers (Baines et al., 2024).

Improved Learning Outcomes Through Data-Driven Insights

AI tools generate performance analytics that help organizations measure the impact of training, identify high-performing learners, and forecast future workforce needs. These insights support more strategic decisions around promotions, succession planning, and skill development investments (Ulfert et al., 2024).

Accelerated Onboarding and Compliance Training

AI streamlines onboarding by offering tailored, role-specific content to new hires, improving time-to-productivity. It also supports compliance training by tracking completion rates and surfacing areas where knowledge may still be lacking (Asfahani, 2022).

Support for Strategic Workforce Planning

AI does not just enhance training; it also helps leaders think ahead. Platforms that integrate workforce analytics and learning systems provide organizations with a real-time view of internal talent capabilities, helping HR teams match training with long-term strategic goals (Baines et al., 2024).

Example Applied AI Tools

A range of AI-powered tools are now being used to enhance corporate learning environments, offering more tailored and interactive training experiences. The examples below illustrate how organizations are applying these technologies to personalize learning, track progress, and improve skill development at scale.

EdCast by Cornerstone
- **What it does**: An AI-powered learning experience platform (LXP) that personalizes training recommendations based on role, interests, and behavior.
- **In this context**: Supports employee reskilling and upskilling by curating internal and external content into individualized learning paths.
- **Evidence**: Used by companies like Dell and Schneider Electric to improve learner engagement and reduce skill gaps (Baines et al., 2024).

Docebo Learn
- **What it does**: AI-enhanced learning management system that personalizes content, automates tagging, and provides predictive analytics on learner performance.
- **In this context**: Helps learning and development (L&D) teams track training effectiveness and identify at-risk learners early.
- **Evidence**: Bankins et al. (2023) describe its use in scalable, data-driven enterprise training programs.

Talespin
- **What it does**: Uses AI and virtual reality to deliver immersive, scenario-based learning experiences.
- **In this context**: Builds soft skills such as communication, leadership, and conflict resolution through simulated practice environments.
- **Evidence**: Asfahani (2022) highlights its role in experiential learning for skill retention and confidence building.

Sana Labs
- **What it does**: Adaptive learning platform that uses AI to continuously personalize training content based on learner performance.
- **In this context**: Enables organizations to tailor content pacing and complexity to individual learning curves.

- **Evidence**: Cited by Ulfert et al. (2024) for its impact on training retention and learner satisfaction across healthcare and education sectors.

ChatGPT for Corporate Training
- **What it does**: Serves as a conversational assistant for explaining complex topics, answering FAQs, or simulating coaching interactions.
- **In this context**: Enhances accessibility to real-time support and reinforces learning outside formal training modules.
- **Evidence**: Emerging use cases documented in Bankins et al. (2023) for knowledge reinforcement and onboarding.

Practical Example: AI-Driven Training at Johnson & Johnson

Background
In response to the fast-changing digital landscape, Johnson & Johnson (J&J) identified the need to upskill its workforce more efficiently. Traditional training methods were often too slow and failed to meet individual learning needs. To address this, J&J integrated AI into its training strategy (Staton, 2024).

Implementation
- **AI-Powered Skill Assessment**: J&J implemented an AI system known as "skills inference" to analyze current employee capabilities and identify skill gaps.
- **Customized Learning Paths**: Based on this analysis, employees received personalized training recommendations aligned with their roles and development goals.
- **Ongoing Skill Development**: The AI-supported system continually tracks workforce capabilities and suggests learning opportunities in real time.

Results
- **Improved Capability Insights**: J&J now has a dynamic view of its workforce's skills, which informs both training and talent planning.
- **Enhanced Learning Efficiency**: AI-enabled learning paths help reduce time spent on irrelevant training.
- **Strategic Workforce Development**: The system has become a critical tool in aligning employee development with organizational needs.

This example highlights how AI can go beyond automation to play a strategic role in workforce development. By embedding intelligence into learning systems, J&J may be better able to build a more agile and future-ready workforce.

Concluding Remarks

AI is beginning to shape how organizations approach employee learning and development, offering the potential for more personalized, scalable, and data-informed training experiences. Tools like adaptive learning systems and virtual coaching platforms can support individual growth while aligning with broader workforce goals.

Additionally, like other AI tools across the employee lifecycle, training platforms depend on transparent, bias-aware implementation to deliver long-term value. Industrial-organizational psychologists play a key role in guiding these efforts, helping ensure that AI-supported learning remains effective, inclusive, and grounded in ethical practice.

AI-Powered Workplace Analytics & Performance Evaluation

As organizations navigate increasingly digital and hybrid work environments, traditional performance management systems are being reexamined. Challenges such as inconsistent evaluation methods, limited feedback cycles, and subjectivity can affect both employee development and organizational effectiveness. AI technologies are being explored as ways to support these systems—offering real-time data, structured feedback, and performance insights that may help improve consistency and responsiveness.

At the same time, these technologies raise important questions about transparency, fairness, and how data is used. Industrial-organizational psychologists and Human Resource (HR) leaders play a key role in shaping how AI is implemented, ensuring that it supports—not overrides—human judgment. This section examines how AI is being used in performance management, with examples from current tools, an example from Indonesia's state-owned postal and logistics company PT. Pos Indonesia, and a look at both the opportunities and challenges of integrating AI into workplace analytics.

Opportunities and Benefits

AI is increasingly being applied to help organizations evaluate performance, track engagement, and manage workforce development in real time. As remote and hybrid work environments challenge traditional evaluation models, AI offers scalable, data-driven alternatives that can improve accuracy, timeliness, and fairness—when implemented with ethical safeguards.

Reducing Subjectivity in Performance Reviews
AI systems rely on quantifiable data—such as task completion, communication behavior, and customer feedback—to produce more consistent evaluations across teams. By minimizing reliance on subjective manager ratings, these tools can support fairer, evidence-based assessments (Baines et al., 2024).

Delivering Real-Time Feedback and Continuous Improvement
AI platforms enable continuous monitoring of employee performance, providing timely, actionable feedback. This supports a culture of immediate response and learning, allowing employees to make adjustments as needed. Over time, this builds the foundation for real-time development—an approach essential for future leadership applications that depend on ongoing behavioral insights and adaptation (Asfahani, 2022). The increased transparency and speed of feedback also strengthen communication between managers and staff, reinforcing continuous improvement practices.

Proactively Identifying Risk and Burnout
Machine learning models can analyze sentiment in emails, survey responses, or collaboration tools to flag early signs of disengagement or burnout. These early warnings give HR teams an opportunity to intervene before issues escalate (Bankins et al., 2023).

Streamlining Performance Review Workflows
By automating the collection and analysis of performance data, AI can reduce administrative burden and speed up review cycles. This efficiency allows HR teams to reallocate time toward coaching, training, and strategic talent planning (Dwianto et al., 2024).

Enhancing Transparency Through Explainable AI
AI platforms equipped with explainability features can show how specific behaviors

or outcomes affect performance scores. This transparency helps employees understand how decisions are made and builds trust in the evaluation process. When employees can see the connection between their actions and their ratings, they're better positioned to understand and improve their development path (Baines et al., 2024).

Scaling Performance Management Across Distributed Teams
In global and hybrid organizations, AI enables a standardized approach to evaluation across roles and locations. This scalability supports fairness in large or fast-growing workforces and ensures consistency in promotion and development decisions (Bankins et al., 2023).

Supporting Data-Driven Workforce Planning
Aggregated performance analytics provide leaders with insight into organizational trends, such as emerging skill gaps or areas for restructuring. These insights can inform strategic decisions around training, promotions, or succession planning (Bankins et al., 2023).

Balancing Productivity Monitoring with Ethics
AI can support productivity goals without compromising employee autonomy—when designed with transparency, opt-in policies, and clear boundaries. Ethical implementation is key to maintaining employee trust and avoiding perceptions of surveillance (Dwianto et al., 2024).

Example Applied AI Tools

A growing number of AI-powered platforms are being integrated into performance management systems to provide continuous feedback, improve evaluation accuracy, and identify workforce trends. Below are several examples of tools currently in use across industries, illustrating how AI can assist with employee development, productivity insights, and organizational decision-making when implemented with transparency and care.

Microsoft Viva Insights
- **What it does**: Aggregates data from Microsoft 365 tools (e.g., Outlook, Teams) to generate reports on focus time, collaboration patterns, and meeting overload.
- **In this context**: Helps organizations track productivity trends and identify burnout risk while maintaining user privacy through aggregated, de-identified data.

- **Evidence**: Used by Accenture to support hybrid workforce well-being and improve meeting efficiency (Baines et al., 2024).

Humanyze
- **What it does**: Uses wearable badges and digital communication data to analyze team dynamics, collaboration frequency, and network efficiency.
- **In this context**: Supports performance evaluation by identifying communication bottlenecks and aligning behavior with business outcomes.
- **Evidence**: Asfahani (2022) highlights its role in identifying productivity trends in large-scale enterprise teams.

Culture Amp
- **What it does**: Combines employee feedback tools with machine learning to analyze engagement, performance, and turnover risk.
- **In this context**: Enables data-driven performance reviews and continuous improvement through real-time feedback and sentiment analysis.
- **Evidence**: Bankins et al. (2023) discuss its use in scaling feedback loops and improving leadership development in mid-size firms.

Betterworks
- **What it does**: AI-enhanced performance management platform that tracks goal alignment, feedback frequency, and progress toward Objectives and Key Results (OKRs).
- **In this context**: Automates performance tracking and provides predictive analytics to support fair and transparent evaluation.
- **Evidence**: Baines et al. (2024) cite its use in replacing static performance reviews with dynamic goal tracking in enterprise organizations.

Synergita Perform
- **What it does:** Leverages AI to analyze appraisal data, generate personalized performance reports, and detect competency gaps.
- **In this context**: Helps managers make more informed decisions and supports employee development planning.
- **Evidence**: Cited in Bankins et al. (2023) as a tool used in manufacturing and information technology (IT) sectors to improve performance calibration.

Practical Example: AI-Based Performance Evaluation at PT. Pos Indonesia

Background

Indonesia's state-owned postal and logistics company, PT. Pos Indonesia implemented an AI-powered performance evaluation system to boost workforce productivity and address limitations in traditional performance reviews. Challenges included subjective assessments, inconsistent feedback, and limited visibility into employee contributions. To modernize these processes, the organization deployed AI-driven analytics within its performance management framework (Dwianto, Kusuma, & Junengsih, 2024).

Implementation

The AI system evaluated a range of metrics to deliver a more objective view of employee performance:

- **Task Completion Rates**: AI analyzed work assignments and deadline adherence to assess efficiency.
- **Customer Feedback**: Machine learning models interpreted customer satisfaction data to measure service quality.
- **Behavioral Patterns:** The system tracked work behaviors to identify strengths, gaps, and potential signs of burnout.

Employees and managers received automated reports, real-time feedback, and personalized performance recommendations tailored to role-specific objectives.

Key Findings

- **Objectivity and Fairness**: The AI system reduced subjective bias, bringing greater consistency and transparency to evaluations.
- **Efficiency Gains**: Automation lightened HR workloads and allowed more focus on employee development.
- **Trust and Ethical Concerns**: Employees appreciated the system's accuracy but raised valid concerns around data privacy, algorithmic bias, and transparency, highlighting the need for ongoing governance and human oversight (Dwianto et al., 2024).

This case illustrates the potential of AI to streamline performance evaluation while reinforcing that fairness, trust, and clear communication must underpin any tech-driven HR transformation.

Concluding Remarks

AI-driven performance evaluation systems offer new ways to increase consistency, provide timely feedback, and support workforce development. When thoughtfully applied, these tools can help organizations identify strengths, address challenges early, and align employee growth with business goals.

At the same time, concerns around bias, transparency, and employee trust remain important. As with other workplace uses of AI, performance systems must be guided by ethical safeguards and human oversight to maintain trust and fairness. Industrial-organizational psychologists have a key role to play in shaping responsible use, helping organizations navigate innovation while staying grounded in fairness and accountability.

AI in Employee Well-being & Mental Health Monitoring

Employee mental health has become a critical factor in workplace performance, retention, and morale; however, traditional methods of monitoring well-being, such as annual surveys or reactive interventions, often fail to address the immediate needs of today's workforce. AI offers promising potential to provide more timely, personalized, and scalable mental health support.

This section examines how AI can help transition from reactive wellness programs to more proactive, data-driven approaches. It highlights practical tools, shares an example from behavioral health, and discusses the emerging benefits of AI-supported well-being systems in the workplace.

Opportunities and Benefits

AI technologies are opening new possibilities for how organizations approach employee mental health, shifting from reactive, one-size-fits-all programs to proactive, personalized, and data-informed systems. When implemented thoughtfully, these tools can offer timely support, improve care access, and generate actionable insights for promoting well-being at scale.

Real-Time Monitoring and Early Intervention

Traditional wellness surveys often come too late to catch rising stress. AI can analyze communication data, behavioral patterns, or biometric signals in real time to flag early signs of burnout, disengagement, or overload, allowing HR teams to intervene before issues escalate (Baines et al., 2024).

Reducing Subjectivity and Enhancing Accuracy

Mental health is difficult to assess through self-report alone, especially in workplace settings. AI can supplement subjective data with sentiment analysis, voice tone recognition, and behavioral analytics, providing a more objective and nuanced view of well-being (Bankins et al., 2023).

Personalized and Adaptive Support

AI platforms can tailor interventions—such as coaching prompts, resilience tools, or cognitive behavioral exercises—based on individual stress profiles, work habits, and engagement history. These personalized experiences increase relevance and improve user outcomes (Ulfert et al., 2024).

Improved Access to Mental Health Resources

AI chatbots and virtual assistant like Wysa offers 24/7 support, expanding access to emotional care, especially in remote or high-stigma environments. This always-on accessibility reduces barriers to help-seeking and complements employee assistance programs (Bankins et al., 2023).

Predictive Burnout Detection

AI models can detect patterns that typically precede burnout—such as declining productivity, missed deadlines, or increased after-hours work—enabling proactive responses. Baines et al. (2024) found that timely AI-driven alerts helped organizations reduce stress-related absenteeism.

Scalable Insights for Organizational Planning

Aggregated, anonymized data allows leaders to identify broader patterns, such as teams at risk of burnout or departments with low morale. These insights can inform systemic changes to workloads, policies, or communication norms (Asfahani, 2022).

Strengthening Trust Through Ethical Monitoring

AI systems that prioritize transparency, opt-in consent, and data anonymization can balance the need for insight with the protection of employee autonomy. Clear communication around how data is used helps maintain trust and reinforces a culture of psychological safety (Dwianto et al., 2024).

Enhanced Culture and Retention

Employees who feel seen and supported in their mental health are more likely to stay and contribute meaningfully. AI tools that facilitate continuous well-being check-ins and personalized support can improve overall engagement and foster a more empathetic work culture (Bankins et al., 2023).

Example Applied AI Tools

The following tools showcase how AI is currently being applied to monitor mental health, support employee well-being, and guide organizational responses:

Microsoft Viva Insights

- **What it does**: As introduced in the Performance Evaluation section of this chapter, Microsoft Viva Insights also supports proactive mental health strategies by monitoring work patterns—such as collaboration load and focus time—to help flag early signs of burnout.
- **In this context**: Supports proactive mental health strategies by flagging risks like meeting overload or lack of recovery time.
- **Evidence**: Baines et al. (2024) highlight its use in organizations aiming to reduce digital burnout in hybrid and remote workforces.

Wysa

- **What it does**: First introduced in the context of employee engagement, Wysa also plays a key role in workplace well-being by delivering confidential, AI-guided emotional support and resilience tools.
- **In this context**: Offers personalized mental health coaching while maintaining user privacy.
- **Evidence**: Ulfert et al. (2024) cite its use in scaling mental health access across large, distributed workforces.

Eleos Health

- **What it does**: Uses AI to analyze therapy session transcripts and voice recordings to detect emotional tone, stress markers, and engagement cues.
- **In this context**: Originally built for clinicians, its capabilities are being explored for organizational wellness programs and employee support services.
- **Evidence**: Demonstrated in practice by Dr. Lior Biran to enhance real-time therapy insights and planning (Eleos Health, 2021).

Koa Health

- **What it does**: Combines AI-driven behavioral analytics with clinically validated digital mental health content.
- **In this context**: Helps employers identify population-level mental health trends and tailor digital interventions accordingly.
- **Evidence**: Asfahani (2022) describes its use in enterprise mental health strategy for proactive support and Return on Investment (ROI) tracking.

Practical Example: Microsoft's Use of AI to Support Workforce Well-being

Background

As remote and hybrid work became widespread, Microsoft recognized growing concerns about digital overload, constant connectivity, and declining employee focus. Traditional well-being surveys proved too slow to catch real-time stressors, prompting the need for a more adaptive approach.

Implementation

Microsoft deployed Viva Insights, an AI-powered platform integrated into Teams and Outlook, to monitor and analyze aggregated, anonymized work patterns. Features include:

- **Collaboration Load Tracking** – AI analyzes meeting time, after-hours work, and email volume.
- **Focus Time Recommendations** – Employees receive nudges to block time for deep work and recovery.
- **Team Insights** – Managers gain visibility into group-level well-being trends without accessing individual data.

Key Results

- **Proactive Burnout Detection:** Organizations identified work patterns linked to stress before they resulted in burnout.
- **Healthier Work Practices:** Many teams adopted protected focus time and reduced meeting overload.
- **Scalable Support**: AI enabled continuous well-being monitoring across thousands of employees, regardless of location.

Conclusion

Microsoft's implementation of AI through Viva Insights illustrates how organizations can move from reactive wellness programs to data-informed, real-time mental health support at scale. It aligns with key challenges in employee well-being: timeliness, personalization, and privacy-centered monitoring (Bankins et al., 2023).

Concluding Remarks

As organizations confront growing demands for mental health support, AI technologies offer valuable opportunities to scale care, enhance personalization, and enable earlier intervention; however, achieving these benefits requires maintaining trust, transparency, and ethical deployment. Industrial-organizational psychologists play a key role in ensuring AI is implemented in a way that respects privacy, supports autonomy, and complements human-led mental health strategies. Going forward, both research and practice should prioritize fairness, interpretability, and inclusivity in AI-powered well-being systems.

———————— ◾● ————————

AI-Powered Decision Support for Leadership & Organizational Strategy

Effective leadership requires not only instinct but also clarity, speed, and the ability to navigate complex and sometimes ambiguous information. Traditional decision-making processes, however, can be impacted by cognitive bias, fragmented data, and slow response times. AI-powered decision support systems offer potential benefits by leveraging predictive analytics, natural language processing, and machine learning. These tools can assist leaders in identifying risks earlier, simulating potential outcomes, and aligning workforce strategies with organizational goals. This section examines how AI can support leadership and strategic planning, highlighting both real-world applications and evidence-based benefits.

Opportunities and Benefits

AI technologies offer the potential to revolutionize how organizations approach performance evaluation and strategic leadership decisions. By shifting from static reviews to dynamic, data-driven insights, AI can enhance objectivity, improve decision-making, and provide real-time visibility into workforce dynamics. Below are several key areas where AI can significantly support leadership and performance evaluation, alongside the benefits it brings to organizational strategy and employee development.

Improved Objectivity and Consistency in Performance Evaluations
Traditional performance reviews are often subjective, relying heavily on managers' perceptions, which can vary widely across teams. AI reduces this subjectivity by analyzing quantifiable data—such as task completion, customer feedback, and collaboration metrics—to offer more objective and consistent performance assessments. Baines et al. (2024) report that this data-driven approach has led to greater fairness and transparency across departments.

Real-Time, Continuous Feedback
Expanding on earlier use in training and evaluation contexts, AI supports leadership with real-time feedback loops that foster proactive coaching and timely course correction. This dynamic feedback loop fosters greater engagement and accelerated development as employees receive timely guidance tailored to their performance. Asfahani (2022) found that organizations using AI-driven performance tools saw quicker resolution of issues and improved communication between employees and managers.

Early Detection of Engagement or Burnout
AI tools can analyze behavioral patterns and sentiment across multiple communication channels to detect signs of disengagement or burnout before they escalate. By monitoring shifts in productivity, communication tone, or work patterns, AI can alert HR teams to potential risks, enabling early intervention. Bankins et al. (2023) highlighted that AI-assisted tools helped reduce burnout-related absenteeism through proactive support.

Streamlined Performance Review Workflows
AI automates the collection and analysis of performance data, freeing HR teams from manual data entry and administrative tasks. This efficiency allows HR teams to focus on higher-value activities, such as talent development and strategic workforce planning. Baines et al. (2024) noted that AI systems helped organizations

streamline their performance review processes, saving time and improving decision-making speed.

Support for Strategic Talent Development and Succession Planning

AI tools can identify high-potential talent by analyzing career trajectories, performance data, and skill sets, ensuring that leadership development is based on objective insights rather than informal networks or subjective opinions. Ulfert et al. (2024) found that AI-powered talent identification led to more equitable development opportunities and higher internal mobility.

Better Alignment of People and Organizational Strategy

AI enables HR and leadership teams to align workforce decisions with broader organizational goals. Real-time data insights help organizations adjust quickly to changes, ensuring that talent strategies remain agile and aligned with business priorities. Baines et al. (2024) report that AI-driven workforce planning tools improved alignment between talent management and business objectives, leading to better long-term outcomes.

Faster, More Informed Decision-Making

AI systems can analyze vast amounts of data to simulate various strategic scenarios and provide decision-makers with predictive insights on potential outcomes. This helps leaders make informed choices quickly, reducing decision-cycle time and improving organizational agility. Bankins et al. (2023) found that AI-powered tools helped reduce the time required to generate insights, allowing for faster, more efficient decision-making.

Reducing Cognitive and Systemic Bias

AI systems can mitigate unconscious bias by relying on structured data and analytics rather than subjective judgment. By grounding leadership assessments in measurable trends, AI ensures that decisions are based on objective evidence, promoting fairness and diversity. Asfahani (2022) found that AI-supported assessments resulted in more diverse leadership pipelines across multiple sectors.

Proactive Risk Identification and Organizational Health Monitoring

AI can continuously monitor organizational health by aggregating data on performance, engagement, and turnover risk. This real-time visibility allows leaders to identify emerging trends or risks—such as declining morale or increasing attrition—before they impact business outcomes. Asfahani (2022) reports that AI helped organizations adjust their workforce strategies in response to real-time insights, improving overall organizational health.

Better Leadership Selection and Development

AI tools can support leadership selection by evaluating a broad range of performance metrics, career progression, and behavioral data. This approach helps identify high-potential candidates for leadership roles based on objective data rather than informal networks or subjective biases. Kiron et al. (2022) discussed how AI tools in leadership development at IBM helped identify emerging leaders more fairly and quickly.

Example Applied AI Tools

These tools illustrate how AI is being used to support data-informed leadership decisions, workforce planning, and organizational agility.

IBM Watson Career Coach
- **What it does**: Offers AI-powered career and leadership development guidance by analyzing employee profiles, skills, and goals.
- **In this context**: Assists HR and leadership teams in matching employees to internal mobility opportunities and succession tracks.
- **Evidence**: Deployed at IBM to enhance workforce planning and reduce attrition by aligning employee development with company needs (Kiron & Spindel, 2019).

Visier People
- **What it does**: A workforce analytics platform that uses AI to identify trends in attrition, performance, and employee engagement.
- **In this context**: Provides leadership with actionable insights on workforce risk and strategic planning across departments.
- **Evidence**: Used by enterprises to inform diversity strategies and leadership pipeline development (Baines et al., 2024).

Eightfold Talent Intelligence Platform
- **What it does**: Uses AI to map skills, recommend training, and identify leadership potential based on both internal data and global labor market trends.
- **In this context**: Helps organizations make fairer and faster talent decisions aligned with business strategy.
- **Evidence**: Bankins et al. (2023) cite its use in enterprise-level workforce transformation projects.

Workday Prism Analytics
- **What it does**: Merges financial, HR, and operational data to offer unified analytics dashboards with embedded machine learning.
- **In this context**: Supports leaders in aligning headcount planning, compensation modeling, and organizational restructuring.
- **Evidence**: Asfahani (2022) discusses its role in streamlining workforce analytics for strategic decision-making.

Sage People
- **What it does**: Uses AI to automate leadership reporting, performance forecasting, and team sentiment analysis.
- **In this context**: Enables mid-size companies to gain enterprise-level insight without requiring a dedicated analytics team.
- **Evidence**: Ulfert et al. (2024) highlight its impact in boosting leadership responsiveness through real-time organizational insights.

Practical Example: IBM's Strategic HR Transformation through AI and Ecosystem Thinking

Background
As the nature of work evolved, IBM recognized that traditional HR approaches were no longer sufficient to manage a dynamic, tech-driven workforce. Rather than focusing solely on full-time employees, IBM began viewing its talent strategy through the lens of workforce ecosystems, integrating AI, analytics, and partnerships to manage a broader range of contributors. This shift aimed to improve decision-making, talent mobility, and workforce agility (Kiron, Spindel, Unruh, & Hancock, 2022).

Implementation
IBM deployed a range of AI-powered strategies to align HR practices with its evolving workforce model:

- **AI-Enabled Talent Management:** IBM used analytics to map employee skills and recommend personalized career paths, helping internal talent navigate opportunities across the enterprise.
- **Workforce Planning at Scale:** AI and data-driven tools enabled real-time insights into workforce composition, project demands, and talent gaps across internal and external contributors.

- **Leadership and Learning Ecosystems:** IBM integrated AI into leadership development, supporting the identification of high-potential individuals and connecting them to targeted learning experiences.

Key Results

- **Ecosystem Agility:** IBM gained the ability to manage a dynamic mix of employees, freelancers, and partners through centralized, AI-informed insights.
- **Enhanced Talent Mobility:** AI-guided career navigation helped increase internal movement and reduce turnover.
- **Data-Driven Decision-Making:** HR and business leaders were better equipped to make workforce decisions grounded in real-time analytics rather than intuition.

Conclusion

IBM's transformation shows how AI can be a strategic enabler—not just a tool—when paired with a broader shift toward workforce ecosystems. This model supports more responsive, inclusive, and data-informed talent practices that go beyond traditional organizational boundaries (Kiron et al., 2022).

Concluding Remarks

When applied thoughtfully, AI-powered decision support systems can provide leaders with enhanced visibility, faster insights, and more consistent decision-making across the organization; however, to fully realize their potential, it is crucial that these systems are transparent, ethically sound, and balanced with human judgment. Industrial-Organizational Psychologists play a critical role in guiding these efforts, integrating behavioral insights with technological advancements to foster leadership that is both strategic and human-centered.

Conclusion

AI is becoming an influential force in Organizational Psychology, offering tools that streamline hiring, assess leadership, and monitor workforce well-being. These systems can boost efficiency and support proactive interventions, but only when used thoughtfully.

We now turn to a different context where AI is influencing care and learning environments: Educational Psychology.

Chapter 7: AI in School & Educational Mental Health Practice

In this chapter, we explore the transformative potential of artificial intelligence (AI) in the field of School and Educational Psychology. AI offers new opportunities for enhancing student support by enabling personalized, data-driven tools that address both academic and emotional challenges. By automating administrative tasks, identifying learning differences earlier, and customizing interventions for individual needs, AI is helping educational professionals create more effective and efficient practices.

In response to challenges of traditional education systems—such as resource shortages and limited teacher availability—AI tools offer the potential to provide scalable solutions that benefit both students and educators. This chapter dives into some ways AI is changing the landscape of School and Educational mental health, from enhancing early diagnosis and intervention to streamlining administrative workloads.

Note: The reader is also referred to the AI in Neuropsychology section of Chapter 8: AI in Health Psychology for innovations involving cognitive and neuropsychological assessments.

Personalized Learning & Student Performance Prediction

Many educational systems rely on standardized models that assume students learn in similar ways and at similar paces. These one-size-fits-all approaches often fail to meet the diverse cognitive, emotional, and academic needs found in real classrooms. Students with learning differences, attention challenges, or variable home support may struggle to keep up. Traditional assessments may be delayed and inflexible, making it difficult to detect problems early or adjust recommended instruction and needed supports in time. Educators are also limited by time and resources, reducing their ability to provide individualized support.

Opportunities and Benefits

Artificial intelligence can help address these challenges by enabling more personalized, timely, and data-informed approaches to learning. AI-driven platforms offer real-time insights, customized learning paths, and predictive models that help identify and support students before they fall behind.

Creating Adaptive, Student-Centered Learning Paths

AI systems customize instruction in real time based on each student's progress, learning style, and knowledge gaps. This allows schools to move beyond rigid, one-size-fits-all curricula. Tools such as Squirrel AI and DreamBox Learning dynamically tailor content, helping students master material at their own pace (Crompton & Burke, 2023; Matochová & Kowaliková, 2024).

Early Detection of At-Risk Students

AI can identify students at risk of falling behind earlier than traditional assessments. By analyzing academic data, engagement trends, and performance patterns, AI tools flag students who may need targeted support. Lin and Chen (2024) found that these systems predicted academic struggles with more than 85% accuracy, allowing for faster intervention.

Real-Time Feedback and Instructional Adjustment

AI-powered platforms such as Carnegie Learning MATHia and Amira Learning give students immediate feedback during learning tasks. This helps correct misconceptions early, supports ongoing skill development, and reduces the delays associated with traditional grading and review cycles (Lin & Chen, 2024).

Scalable Personalization for Diverse Classrooms

AI enables personalized instruction at scale, even in classrooms with high student-to-teacher ratios. Platforms like Knewton Alta allow teachers to support a wide range of learners simultaneously by automatically adjusting content difficulty, pacing, and review materials based on each student's performance (Klimova & Pikhart, 2025).

Data-Driven Decision-Making for Instructional Support

AI systems synthesize data from quizzes, participation, and behavioral trends into clear, actionable insights for educators. These insights inform lesson planning and instructional strategies, helping teachers respond quickly to student needs (Crompton & Burke, 2023).

Improved Equity in Assessment and Learning

By analyzing multiple data sources rather than relying solely on standardized tests, AI platforms can reduce cultural, linguistic, and learning-style biases in student evaluations. This supports more accurate and inclusive assessments, especially for historically underserved populations (Klimova & Pikhart, 2025).

Improved Outcomes Through Timely, Tailored Intervention

AI helps align instructional content with each student's current level of understanding. By adjusting learning pathways in real time, AI platforms reduce both under- and over-challenging students. This improves academic outcomes and supports steady progression across subjects (Crompton & Burke, 2023).

Example Applied AI Tools

This section highlights real-world AI tools designed to personalize learning, support early intervention, and provide scalable academic feedback. These examples show how AI can help address key instructional challenges by supporting individual student needs in real time.

Squirrel AI

- **What it does**: A Chinese adaptive learning platform that uses AI to personalize instruction based on students' mastery of micro-concepts in real time.
- **In this context**: Supports tailored remediation and learning acceleration, particularly effective for large-scale classroom settings.

- **Evidence**: Cited by Crompton and Burke (2023) for its role in adaptive tutoring and early performance prediction.

Carnegie Learning MATHia
- **What it does**: AI-powered math tutoring software that provides step-by-step support and tracks student learning at a granular level.
- **In this context**: Offers real-time feedback, identifies misconceptions, and customizes content to student progress.
- **Evidence**: Documented improvements in middle and high school math performance in U.S. districts (Lin & Chen, 2024).

DreamBox Learning
- **What it does**: Adaptive K–8 math platform that adjusts in real time to students' responses and learning pace.
- **In this context**: Helps differentiate instruction and personalize content for diverse learners.
- **Evidence**: Matochová & Kowaliková (2024) support the use of AI-driven adaptive learning systems to enhance educational equity, particularly for underserved student populations.

Knewton Alta
- **What it does**: Adaptive learning system for higher education that uses AI to personalize digital coursework based on student performance.
- **In this context**: Enables early detection of learning gaps and provides customized review materials.
- **Evidence**: Klimova & Pikhart (2025) highlight the role of adaptive AI in improving student engagement, reducing stress, and enhancing content retention in university settings.

Amira Learning
- **What it does**: AI-powered reading tutor that listens to students read aloud and provides fluency assessments and feedback.
- **In this context**: Assists early literacy development and supports individualized reading interventions.
- **Evidence**: Recognized in Lin & Chen (2024) for use in early education to identify struggling readers.

Practical Example: AI in Student Performance Prediction

Background

Artificial intelligence (AI) is increasingly being used in education to identify students at risk of underperformance and to tailor academic support in real time. Lin and Chen (2024) explored how AI-integrated educational applications can predict student engagement and academic struggles by analyzing behavioral patterns, online activity, and academic performance. Similarly, Crompton and Burke (2023) identified predictive modeling as a key application of AI in higher education, helping institutions allocate resources more effectively and intervene earlier in the learning process.

Implementation

AI-powered systems in these studies utilized a combination of behavioral data and machine learning techniques to forecast academic outcomes.

- **Performance Analytics:** Student interaction data—such as time spent on tasks, quiz scores, and participation—was fed into predictive algorithms to assess learning trends.
- **Engagement Monitoring:** The AI monitored real-time engagement and flagged early warning signs of disengagement or cognitive overload.
- **Personalized Recommendations:** Based on predictive insights, the systems offered individualized content adjustments and alerts to educators for timely intervention.

Key Findings

- **Improved Student Engagement:** Gamified features and personalized feedback loops helped boost motivation and reduce emotional disengagement (Lin & Chen, 2024).
- **Early Identification of Learning Gaps:** Predictive models accurately identified students at risk of academic decline, enabling proactive support (Crompton & Burke, 2023).
- **Customized Learning Paths:** AI dynamically adapted content delivery based on individual performance trends and learning preferences (Lin & Chen, 2024).
- **Data-Informed Decision-Making:** Institutions gained a clearer picture of student needs, allowing for more strategic allocation of resources and targeted interventions (Crompton & Burke, 2023).

Conclusion

These case studies demonstrate how AI enables a shift from reactive to proactive academic support by continuously monitoring student behavior, predicting challenges, and personalizing learning at scale. While results show promising gains in engagement and retention, the studies also underscore the importance of ethical implementation, transparency, and maintaining a human role in interpreting AI-generated insights.

Concluding Remarks

Artificial intelligence (AI) is reshaping how school and educational psychologists address core academic challenges by enabling more personalized, data-informed interventions. Adaptive learning systems and predictive analytics allow for earlier identification of students at risk, real-time instructional adjustments, and scalable support tailored to individual needs. These tools offer new ways to move beyond standardized teaching and reach students who may otherwise fall behind; however, to ensure these systems serve all learners equitably, transparency, data privacy, and educator oversight must remain central. AI should enhance—not replace—the human relationships and professional judgment that are essential to meaningful education.

AI in Special Education & Inclusivity

Students with disabilities and neurodivergent profiles often encounter barriers to learning in environments built around standardized instruction. Traditional education systems may lack the flexibility to meet individual needs or the tools to make content fully accessible to students with speech, hearing, or language impairments. These limitations can create persistent gaps in participation, progress, and inclusion.

Opportunities and Benefits

Artificial intelligence (AI) can help address these challenges by enabling more tailored instructional support and improving access to educational content. Through adaptive learning platforms, assistive communication technologies, and AI-driven personalization, these tools offer scalable solutions to better serve students with varying abilities and neurodivergent needs. Below are a few ways AI can support more inclusive and individualized learning experiences.

Adaptive Instruction for Diverse Learning Profiles

AI-powered learning systems dynamically adjust instruction based on each student's strengths, challenges, and pace. This personalized approach helps students with Dyslexia, Attention Deficit-Hyperactivity Disorder (ADHD), and other neurodivergent conditions engage with content in ways that match their learning profiles. Platforms such as Squirrel AI and DreamBox Learning demonstrate how adaptive instruction can replace one-size-fits-all methods in both general and special education settings (Crompton & Burke, 2023; Matochová & Kowaliková, 2024).

Improved Accessibility for Students with Communication Barriers

AI tools such as text-to-speech, real-time translation, and speech recognition make instructional content more accessible for students with speech, hearing, or language challenges. Tools like Read&Write, Microsoft Translator, and Project Euphonia help remove communication barriers and foster inclusion in both instruction and assessment (Shirley & Nair, 2023; Klimova & Pikhart, 2025).

Example Applied AI Tools

This section highlights some of the key AI tools currently being used to support students with diverse learning needs, particularly in special education. These tools leverage AI to provide personalized, accessible, and adaptive learning experiences, enhancing engagement and promoting inclusion for students with varying learning needs.

Read&Write by Texthelp
- **What it does**: A literacy support tool that offers text-to-speech, word prediction, and audio reading features.
- **In this context**: Assists students with Dyslexia, ADHD, or visual impairments by converting digital text into speech, helping them engage more fully with reading assignments.
- **Evidence**: Used in inclusive classrooms across North America to support reading fluency and comprehension (Shirley & Nair, 2023).

Project Euphonia (Google)

- **What it does**: AI-powered speech recognition model trained to understand diverse and atypical speech patterns.
- **In this context:** Helps students with speech impairments or neurological conditions use voice interfaces and communication devices more effectively.
- **Evidence**: Recognized in accessibility research for improving assistive voice technology (Klimova & Pikhart, 2025).

Amira Learning

- **What it does**: An AI-driven reading tutor that listens to students read aloud and provides fluency assessments and corrective feedback.
- **In this context**: Supports early intervention for students with Dyslexia or other reading disorders by delivering individualized literacy coaching.
- **Evidence**: Featured in Lin & Chen (2024) as improving reading outcomes in special education settings.

Microsoft Translator & Live Captions

- **What it does**: Provides real-time translation and captioning across multiple languages and speech formats.
- **In this context**: Enhances accessibility for students who are deaf, hard of hearing, or English language learners by offering multilingual and multimodal support.
- **Evidence**: Widely adopted in public education systems for inclusive communication (Crompton & Burke, 2023).

Practical Example: AI-Driven Reading Support for Students with Dyslexia

Background

In a private special education setting in Chennai, India, researchers Shirley and Nair (2023) studied the impact of Microsoft's Immersive Reader, an AI-powered tool, on students with Dyslexia. The intervention addressed the limitations of traditional one-on-one literacy support, which can be time-intensive and not scalable. The researchers aimed to test whether AI could deliver personalized, engaging, and effective reading support for students with diverse needs.

Implementation

Eighteen students in grades 4 and 5 with diagnosed Dyslexia participated in the intervention. Over the course of one month, the experimental group used Microsoft Immersive Reader for 30 minutes daily. This AI-driven tool uses speech recognition and natural language processing (NLP) to support reading development through:

- Real-time text-to-speech narration with word highlighting.
- Adjustable pace, font, spacing, and background to improve readability.
- Visual grammar tools (e.g., noun/verb color coding), translation, and syllabification features.
- Personalized voice settings for narration.

Students navigated the tool independently with minimal teacher input, demonstrating increasing confidence and enthusiasm for reading activities.

Key Results

- **Improved Reading Fluency:** Students showed measurable gains in word recognition, fluency, spelling, decoding, and phonological processing based on pre- and post-test scores.
- **Engagement and Motivation:** The tool's multisensory features—including adjustable visuals and real-time auditory feedback—boosted student motivation and reduced reading-related anxiety.
- **Reduced Teacher Workload:** Educators reported being able to shift focus from intensive one-on-one interventions to broader learning support, as the AI tool provided individualized, real-time assistance.
- **Accessibility and Independence:** Students were able to customize the interface to suit their comfort, promoting autonomy in learning.

Conclusion

This study highlights how AI-powered reading tools like Immersive Reader can significantly improve reading skills and learning engagement in students with Dyslexia. While human oversight remains essential, AI can serve as a scalable supplement to traditional instruction, particularly in resource-limited educational settings (Shirley & Nair, 2023).

Concluding Remarks

Artificial intelligence (AI) is helping reshape special education by offering more tailored and accessible learning experiences for students with disabilities and neurodivergent needs. Tools that personalize instruction and reduce communication barriers are creating new opportunities for inclusion in classrooms that have long struggled to meet students with diverse learning profiles.

While these technologies are not a substitute for professional judgment or individualized human care, they can extend the reach and precision of support services. As these systems continue to evolve, thoughtful implementation—with strong safeguards around data use, accessibility, and equity—will be essential. When used responsibly, AI holds promise to help close longstanding gaps in educational access and participation for students who have too often been overlooked by one-size-fits-all approaches.

Addressing Educational Inequality

Persistent disparities in education—shaped by socioeconomic status, geographic isolation, and uneven access to resources—continue to limit opportunities for many students. In low-income and rural areas, learners often face large class sizes, limited tutoring options, and a lack of qualified educators or updated materials. The digital divide compounds these issues, making it harder for students in under-resourced communities to stay engaged and connected in increasingly technology-driven learning environments.

Opportunities and Benefits

Artificial intelligence (AI) offers practical tools to help reduce these inequities. AI-driven platforms can deliver affordable, scalable academic support that adapts to individual needs and can operate across diverse contexts. Especially in settings where human and material resources are scarce, these technologies can extend learning opportunities that might otherwise be out of reach.

Delivering Personalized Learning at Scale
AI-powered platforms such as DreamBox and Khan Academy use adaptive algorithms to tailor lessons in real time based on student performance. This allows learners to progress at their own pace and receive content suited to their current level of understanding. For students in overcrowded or under-resourced classrooms, where individualized attention is limited, this kind of scalable personalization can help close achievement gaps and improve learning outcomes (Crompton & Burke, 2023).

Expanding Access Through Mobile and Offline-Compatible Tools

Many AI applications are designed to function on mobile devices and with limited internet connectivity. This makes them well suited for rural or low-income communities, where reliable broadband access may not be available. These tools can bring educational content to learners who would otherwise face exclusion from digital learning environments due to infrastructure limitations (Lin & Chen, 2024).

Providing Always-On Tutoring and Homework Help

AI-powered tutoring platforms like Socratic by Google offer students academic help whenever they need it. These tools explain concepts, guide problem solving, and support independent learning outside school hours. For students who lack access to private tutoring or at-home academic support, always-on AI tutoring helps level the playing field by offering consistent, low-cost academic assistance (Lin & Chen, 2024).

Example Applied AI Tools

These AI-powered tools are designed to expand access to personalized instruction, offer low-cost academic support, and reach students in under-resourced settings. Each one supports key goals in reducing educational inequality.

Khan Academy with Khanmigo
- **What it does**: Khanmigo is an AI-powered tutor integrated into Khan Academy's platform. It uses personalized prompts and Socratic questioning to guide students through math, science, and writing tasks.
- **In this context**: Offers individualized support to students in low-income or rural schools who may not have access to private tutoring or one-on-one help.
- **Evidence**: Crompton & Burke (2023) highlight Khan Academy's effectiveness in scaling equitable education through adaptive instruction.

Socratic by Google
- **What it does**: A free, AI-powered homework assistant that interprets student questions using OCR and natural language processing, then provides step-by-step solutions.
- **In this context:** Gives students access to academic help anytime, reducing reliance on out-of-school support that may be unavailable in low-resource homes.

- **Evidence**: Lin & Chen (2024) report high engagement and usage among students in hybrid and underserved learning environments.

Microsoft Immersive Reader
- **What it does**: Provides text-to-speech, translation, grammar visualization, and reading assistance through an AI-driven interface.
- **In this context**: Improves literacy access for students in underfunded schools, particularly multilingual learners or those with reading difficulties who need inclusive tools without additional cost.
- **Evidence**: Shirley & Nair (2023) found it improved literacy outcomes in classrooms serving students with limited reading support.

Edmentum Exact Path
- **What it does**: Uses AI to create adaptive learning paths across core subjects, adjusting to student needs based on ongoing diagnostic assessments.
- **In this context**: Helps educators in schools with large class sizes provide targeted instruction, even when one-on-one support is limited.
- **Evidence**: Cited by Matochová & Kowaliková (2024) for its use in high-poverty districts to close learning gaps through differentiation.

Practical Example: Socratic by Google – Expanding Academic Support in Low-Resource Settings

Background
Lin & Chen (2024) examined the use of Socratic by Google as a low-cost academic support tool for students in remote and underserved school settings. In these contexts, many learners lacked consistent access to after-school tutoring or one-on-one help. Socratic offered an accessible, AI-powered assistant to help students navigate assignments independently, even when teacher or caregiver support was limited.

Implementation
Socratic uses natural language processing (NLP) and optical character recognition (OCR) to interpret homework questions and provide step-by-step explanations in subjects like math, science, and literature. Students can scan handwritten or printed problems with their phones, and the app responds with structured, visual guidance tailored to the specific task.

Key Results

- **Widened Access to Help**: Students using Socratic demonstrated improved homework completion rates in remote and hybrid learning environments, especially in districts with limited teacher availability.
- **Reduced Learning Gaps**: Students in low-resource schools who used Socratic were able to maintain progress on par with peers who had access to private tutoring or parental help.
- **High Engagement and Retention:** Students used the platform regularly for homework help, and teachers reported fewer instances of incomplete assignments and increased academic persistence.

Conclusion

This case highlights how free, mobile-friendly AI tools like Socratic can help level the playing field by providing high-quality academic assistance to students in under-resourced educational settings. As part of a broader equity strategy, such tools demonstrate the potential of AI to deliver scalable, just-in-time learning support.

Concluding Remarks

Artificial intelligence offers a practical means to extend quality learning opportunities to students historically underserved by traditional education systems. By lowering barriers to tutoring, adapting instruction at scale, and reaching learners in areas with limited connectivity, AI can help mitigate the structural inequities caused by poverty, geographic isolation, and resource constraints. While these tools cannot replace the broader need for systemic investment in public education, they can complement existing efforts by offering targeted support where human capacity is stretched thin. As these technologies continue to evolve, equity must remain central, ensuring that AI deployment prioritizes inclusivity, accessibility, and the diverse realities of all learners.

Supporting Student Emotional Well-being

Supporting student emotional well-being has become increasingly critical as more students experience stress, anxiety, and disengagement; however, many school systems struggle to provide the timely and adequate support these students need. With counselor-to-student ratios often too low and stigma around seeking help for mental health concerns, emotional well-being frequently takes a backseat in the classroom. Moreover, the absence of real-time emotional monitoring makes it difficult for educators to intervene when students most need support. This section explores how AI-powered tools can help bridge these gaps and provide better emotional and psychological care for students.

Opportunities and Benefits

AI has the potential to address the key challenges schools face in supporting emotional well-being by offering proactive, scalable, and personalized solutions. AI-powered tools can detect signs of emotional distress early, allowing for timely intervention. These systems are particularly valuable in large classrooms or resource-constrained environments, where it may be difficult for educators to monitor every student's emotional state. Furthermore, AI can help reduce the stigma surrounding mental health by providing students with private, accessible avenues for support. By personalizing learning experiences and offering scalable mental health resources, AI can help ensure that emotional well-being is prioritized in educational settings.

Proactive Mental Health Support at Scale

AI-powered mental health apps, such as Wysa and newer tools like Wayhaven, offer 24/7 emotional support for students, providing coping strategies and stress management techniques, especially for those who might not feel comfortable seeking traditional counseling. This helps extend mental health support, particularly where counselors are overburdened.

Early Detection of Emotional Distress

AI can analyze students' written responses, speech tone, and facial expressions to detect early signs of emotional distress, enabling timely intervention. A tool like Wysa offer 24/7 support, while Microsoft Reflect helps educators track emotional trends (Crompton & Burke, 2023; Lin & Chen, 2024).

Enhanced Equity in Emotional Support

AI tools, such as chatbots, offer equitable access to mental health support by reducing the stigma often associated with seeking help. These tools can serve marginalized students who may otherwise hesitate to engage in traditional counseling sessions (Torous et al., 2021).

Reduced Stigma Around Seeking Help
AI interfaces create a private, user-controlled environment where students can check in on their emotional well-being without fear of judgment. This encourages open, honest communication about mental health, helping students feel more comfortable engaging with support resources (Crompton & Burke, 2023).

Example Applied AI Tools

The following AI-powered tools are being utilized or tested in educational settings to provide scalable mental health support, early emotional distress detection, and help reduce the stigma surrounding mental health. Each of these tools showcases how AI can play a role in ensuring that students have the support they need for emotional well-being, whether it is through accessible, real-time assistance or personalized care.

Disclaimer Note:
Please note that some of the AI tools mentioned in this section, specifically the mental health chatbot like Wysa, was previously discussed in Chapter 4: AI in Counseling and Clinical Practices. While these tools are also relevant to supporting students' emotional well-being, their application in educational settings is distinct from their use in clinical contexts, as discussed earlier. This section aims to explore how these tools are specifically adapted to meet the needs of students within school environments.

Wayhaven
- **What it does:** An AI-powered mental wellness companion that offers evidence-informed conversation support based on CBT, DBT, ACT, and mindfulness principles.
- **In this context:** Designed to support student well-being by providing scalable, accessible conversations focused on emotion regulation, distress tolerance, and self-reflection. Especially useful in school settings where mental health resources are limited.
- **Evidence:** Umashankar & Geethanjali (2023) demonstrate that AI-powered chatbots can effectively support college students' mental health by offering immediate, private, and judgment-free interactions. Their findings highlight improvements in emotional resilience, reduced stigma in seeking help, and increased accessibility of care for university populations.

Wysa
- **What it does:** An AI-driven mental health coach that provides mood tracking, guided journaling, and mindfulness exercises.
- **In this context**: Used in schools and universities to support student well-being and emotional resilience.
- **Evidence:** Reported in Matochová & Kowaliková (2024) as improving emotional regulation and daily coping in adolescent learners.

Microsoft Reflect (within Teams for Education)
- **What it does**: A check-in tool that uses simple prompts and AI-supported analysis to help students share how they are feeling.
- **In this context**: Helps educators monitor emotional trends in the classroom, providing a scalable way to track student emotional health and engagement, especially in remote learning environments.
- **Evidence**: Highlighted in Lin & Chen (2024) for its role in monitoring engagement and emotional health.

Practical Example: AI Chatbots and Student Well-being in Higher Education

AI chatbots are increasingly being integrated into university platforms to support student mental health. These systems offer real-time emotional and academic assistance, provide a nonjudgmental space for expression, and help route students to appropriate resources. As Klimova and Pikhart (2025) show, such tools can play a meaningful role in improving emotional well-being in higher education settings.

Key Findings
- **Improved Mental Health Access:** AI chatbots help expand the reach of mental health services, especially for students who may not feel comfortable seeking in-person support. By providing 24/7 emotional and academic assistance, these chatbots become an accessible first line of support, especially in environments where counselor availability is limited.
- **Reduced Academic Stress**: AI systems that assist with workload management, break down complex assignments, and offer time-management guidance have been shown to alleviate academic pressure. This not only reduces stress but also empowers students to manage their academic responsibilities more effectively.
- **Enhanced Emotional Well-being**: Regular interaction with AI tools that offer supportive conversations has been linked to higher levels of emotional

resilience and academic motivation. These tools are particularly beneficial during high-stress times, such as exam periods, helping students feel more grounded and capable of managing stress.

- **Communication Support:** AI-driven applications also facilitate better written communication, enabling students to articulate their thoughts and feelings clearly. This is particularly helpful for students who might struggle with expressing themselves in traditional counseling settings.

These findings demonstrate the potential of AI-powered chatbots to support student emotional well-being by providing a scalable, accessible, and effective solution to mental health challenges, ultimately contributing to both academic and psychological success.

Concluding Remarks

In this section, we discussed how AI-driven tools are showing promise to becoming invaluable in supporting student emotional well-being, particularly in settings where timely intervention is crucial. These technologies provide accessible, real-time emotional support, personalized coping strategies, and early detection of distress. By offering scalable and nonjudgmental mental health resources, AI helps reduce stigma and provides equitable access to emotional support, especially for marginalized students. As AI systems continue to evolve, they hold significant potential to enhance student well-being by improving both academic and psychological outcomes in educational environments.

Boosting Student Engagement

Student engagement is a crucial factor in learning success, yet many students, particularly in large or under-resourced classrooms, struggle with staying engaged. Traditional education systems often fail to provide engaging content or personalized learning experiences that align with students' individual interests and learning styles. The challenges associated with student disengagement are further exacerbated in remote or hybrid learning environments, where students may feel disconnected and unsupported.

Opportunities and Benefits

AI has the potential to revolutionize how educators engage students by providing personalized, interactive, and dynamic learning experiences. AI-powered tools can adapt learning content based on student interests, real-time performance, and emotional responses, keeping students engaged and motivated. Through gamified learning, instant feedback, and tailored learning paths, AI can re-engage students who might otherwise become disinterested or disengaged from the content (Lin & Chen, 2024; Crompton & Burke, 2023).

Emotion-Aware Learning Platforms

AI-powered platforms can adjust lesson difficulty, provide motivational messages, or pause content when stress or frustration is detected. This helps sustain engagement, reduces burnout, and ensures students remain motivated, which can be crucial for those in virtual or overcrowded classrooms (Crompton & Burke, 2023; Matochová & Kowaliková, 2024).

Personalized Learning That Boosts Motivation

AI can personalize learning paths based on individual student progress and emotional states, tailoring content to match their current needs. This not only helps to reduce feelings of frustration and boredom but also boosts students' motivation and academic persistence by providing relevant challenges (Matochová & Kowaliková, 2024).

Scalable Monitoring and Support

AI systems that track student engagement and emotional well-being at scale would allow schools to prioritize interventions effectively. This is especially beneficial in large, under-resourced schools where it is difficult to monitor individual student needs manually (Lin & Chen, 2024).

Example Applied AI Tools

The following AI-powered tools are designed to monitor and enhance student engagement, particularly in environments where traditional approaches may struggle to keep students interested or motivated. By using real-time data and emotional cues, these tools provide personalized experiences, detect disengagement, and offer immediate feedback, creating a more dynamic and interactive learning environment. These examples highlight how AI can help keep students engaged and connected, even in the most challenging learning conditions.

Emotion AI by Affectiva
- **What it does**: An emotion-recognition platform that analyzes facial expressions and voice to assess engagement and emotional states.
- **In this context**: Integrated into online learning tools, it gauges student frustration or confusion and prompts timely adjustments to lesson difficulty or pacing.
- **Evidence**: Cited in Klimova & Pikhart (2025) for its use in emotion-aware tutoring systems.

Replika for Education
- **What it does**: A conversational AI designed to simulate emotionally intelligent dialogue.
- **In this context**: Piloted as a digital companion for students to express feelings, reduce loneliness, and build communication confidence, helping to keep students engaged socially and emotionally during learning.
- **Evidence**: Discussed in early-stage trials as part of school-based social-emotional learning (SEL) research (Garg & Sharma, 2024).

Microsoft Reflect (within Teams for Education)
- **What it does**: A check-in tool that encourages students to share how they feel using AI-supported analysis.
- **In this context**: Used to assess student engagement, especially in hybrid or remote learning settings, and allows educators to respond to emotional disengagement or distress in real-time.
- **Evidence**: Lin & Chen (2024) highlight its ability to monitor engagement and emotional health, ensuring student needs are addressed in digital spaces.

Practical Example: AI-Powered Adaptive Learning in Student Engagement

Student disengagement is a persistent challenge, especially in large or under-resourced classrooms. AI-powered adaptive learning platforms address this by personalizing instruction and adjusting learning paths in real time to match each student's pace and needs (Lin & Chen, 2024).

Implementation
One prominent example is DreamBox Learning, an adaptive K–8 math platform that adjusts in real time to students' responses. DreamBox Learning uses AI algorithms

to personalize lessons based on a student's proficiency and pace, ensuring that each learner is challenged appropriately. The platform can detect when a student is struggling with a concept or is becoming disengaged, adjusting the difficulty of tasks or offering supportive hints to maintain engagement (Crompton & Burke, 2023).

In another case, Squirrel AI, a Chinese adaptive learning platform, provides personalized instruction by tailoring lessons based on a student's mastery of micro-concepts. This real-time adjustment ensures that students receive the appropriate level of challenge, which helps keep them engaged and progressing in the curriculum (Matochová & Kowaliková, 2024).

Key Findings
- **Improved Engagement:** Both DreamBox Learning and Squirrel AI demonstrated improved engagement by adapting lessons to meet individual needs, reducing frustration and boredom for students (Lin & Chen, 2024).
- **Increased Motivation**: Gamified elements in platforms like DreamBox Learning helped boost motivation by providing instant feedback and rewards, which in turn led to higher completion rates for assignments (Crompton & Burke, 2023).
- **Better Learning Outcomes**: By personalizing learning content, AI tools helped students progress at their own pace, leading to improved academic performance, especially in students who had previously been at risk of disengagement (Matochová & Kowaliková, 2024).

Conclusion
These case studies highlight the potential of AI-powered adaptive learning platforms to address disengagement in the classroom. By personalizing content delivery and dynamically adjusting lesson difficulty, these tools help sustain student engagement, making learning more relevant and motivating. The success of platforms like DreamBox Learning and Squirrel AI demonstrates how AI can be used to keep students engaged, even in large, resource-constrained classrooms, and shows that personalized learning approaches are key to maintaining student interest and improving outcomes (Lin & Chen, 2024).

Concluding Remarks

AI-powered tools have proven to be a valuable asset in addressing the challenge of student disengagement, particularly in classrooms where traditional approaches fall short. By leveraging adaptive learning systems that adjust content in real-time based on individual student needs, AI ensures that lessons remain relevant and challenging, keeping students engaged and motivated. The ability to personalize learning paths, provide instant feedback, and offer timely interventions is essential for combating boredom and frustration, especially in remote or overcrowded environments.

As seen in case studies like DreamBox Learning and Squirrel AI, AI platforms can effectively enhance engagement, increase motivation, and improve academic outcomes by catering to each student's unique learning profile. By implementing these technologies, schools can create more dynamic, personalized, and inclusive learning environments that foster sustained student interest and help address disengagement at scale. The integration of AI in education is not just about improving engagement; it is about ensuring that every student receives the individualized support they need to succeed academically, emotionally, and socially.

Automating Administrative Tasks for Educators

Educators, particularly in large or under-resourced schools, often face the challenge of spending a significant portion of their time on administrative tasks such as grading, documentation, and scheduling. These time-consuming duties, while essential, detract from the time available for direct student support, engagement, and teaching. Research indicates that administrative tasks can occupy a substantial part of educators' workdays, leaving little room for high-impact activities that directly benefit student learning and well-being (Crompton & Burke, 2023). The burden of these responsibilities often leads to burnout and reduces the overall effectiveness of educators in fulfilling their primary roles.

Opportunities and Benefits

AI-powered automation directly addresses the key challenges faced by educators and school psychologists by streamlining time-consuming administrative tasks, improving the efficiency and accuracy of documentation, and reducing the burden on overextended staff. By leveraging AI to handle routine tasks, schools can ensure that educators have more time to focus on student engagement, teaching, and support.

Significant Time Savings for Educators and Psychologists

AI helps reduce the substantial amount of time spent on grading, documentation, and scheduling by automating repetitive tasks such as essay grading, attendance logging, and report generation. This results in significant time savings, enabling educators and psychologists to dedicate more hours to high-impact student-focused activities. For instance, AI tools like Gradescope and TARA automate grading and report creation, reducing turnaround times and helping professionals manage their caseloads more effectively (Flodén, 2024; Klimova & Pikhart, 2025).

Improved Accuracy and Consistency in Documentation

Automated systems minimize human error in documentation and compliance tasks, ensuring that records are consistent, accurate, and up-to-date. AI tools like Otus provide automated progress tracking, eliminating inconsistencies often found in manual data entry. This increased accuracy is especially crucial in contexts such as Individualized Education Programs (IEPs), where legal compliance is necessary, ensuring that educators and school psychologists can make data-driven decisions with confidence (Matochová & Kowaliková, 2024).

Scalable Solutions for Large or Understaffed Institutions

Many schools, especially those with limited resources, face staffing shortages that make it difficult to keep up with administrative tasks. AI platforms like ScribeSense and Otus offer scalable solutions that help schools with large student populations or limited staff. These tools automate grading and data analysis, ensuring that even in resource-constrained environments, educators can maintain high-quality services without sacrificing the time needed for student-focused activities (Crompton & Burke, 2023).

Faster, More Actionable Feedback for Students

AI automation accelerates the feedback process, enabling educators to deliver more timely and personalized responses to students. With tools like Gradescope and Socratic by Google, assessments are graded faster, and actionable feedback is provided quickly, allowing teachers to adjust their instructional methods or interventions. This enhanced speed supports self-regulated learning and helps students stay on track with their academic progress (Crompton & Burke, 2023).

Reduced Administrative Burden for School Psychologists

School psychologists often face overwhelming administrative workloads, including writing reports, summarizing testing results, and ensuring compliance with regulations. AI platforms like TARA assist by automating the documentation and report-writing processes, freeing psychologists to focus more on student intervention, collaboration with educators, and direct counseling (Klimova & Pikhart, 2025). These tools support the growing demand for psychological services without compromising the quality of care.

Streamlined Scheduling and Communication

Managing appointments, meetings, and counseling sessions in large schools can be a logistical challenge. AI-integrated tools such as Calendly for Education help streamline scheduling by automatically suggesting optimal meeting times based on availability and school priorities. This reduces scheduling conflicts and improves access to counseling services, ensuring that students receive the support they need when they need it (Lin & Chen, 2024).

Increased Equity in Support Delivery

AI-powered automation helps bridge the gap between well-resourced and under-resourced schools. By automating administrative processes, AI tools allow schools with fewer resources to maintain high standards in grading, feedback, and documentation. This increased efficiency ensures that underserved schools can provide equitable access to timely educational support without overwhelming staff (Matochová & Kowaliková, 2024).

Example Applied AI Tools

A growing number of AI-powered platforms are being implemented to reduce educator workload and streamline school operations. Below are several real-world tools that demonstrate how automation is being used to improve efficiency across administrative workflows, directly addressing the key challenges of excessive time spent on grading, inconsistent manual processes, and resource shortages.

Gradescope

- **What it does**: AI-assisted grading platform that automates the scoring of exams, homework, and coding assignments across multiple disciplines.

- **In this context**: Reduces time spent on grading, improves consistency, and provides item-level analytics for instructional refinement, allowing educators to focus more on teaching and student interaction.
- **Evidence:** Used in universities globally; cited in Flodén (2024) for matching human scoring with up to 70% accuracy while reducing grading time by 30%, providing a more efficient alternative to manual grading.

Otus

- **What it does**: A comprehensive K–12 data platform that automates assessment, attendance, and progress monitoring with AI-backed insights.
- **In this context:** Supports educators by automating reporting and simplifying intervention planning, eliminating the need for manual data entry and reducing administrative load, thus enabling more focus on student engagement.
- **Evidence**: Highlighted in Matochová & Kowaliková (2024) as a scalable tool for automating data analysis, particularly in schools with limited resources.

ScribeSense

- **What it does:** Automates the digitization and scoring of paper-based assessments using AI-powered handwriting recognition.
- **In this context:** Helps schools with limited access to digital resources maintain traditional paper testing while automating scoring and analytics, reducing the burden on educators and addressing resource shortages.
- **Evidence**: Piloted in rural school systems where digital testing is limited; referenced by Crompton & Burke (2023) for reducing administrative lag and minimizing error-prone manual processes.

Calendly for Education (AI-integrated version)

- **What it does**: Uses AI to suggest optimal meeting times based on calendar data, availability, and role-specific priorities.
- **In this context**: Automates scheduling, reducing the administrative burden of organizing student meetings, parent conferences, and team consultations, which is particularly useful in schools facing staff shortages.
- **Evidence**: Featured in Lin & Chen (2024) for improving time management and reducing appointment conflicts, helping school staff spend less time on logistics and more on student-focused activities.

TARA (Task Automation for Remote Assessments)
- **What it does**: AI platform designed for school psychologists to automate cognitive testing logistics, report drafting, and document generation.
- **In this context:** Speeds up testing cycles and reduces documentation burdens for school psychologists, increasing efficiency and enabling them to focus more on providing direct psychological services to students.
- **Evidence**: In pilot use across several districts, cited by Klimova & Pikhart (2025) for increasing capacity for psychological services and reducing time spent on non-student-facing tasks.

Practical Example: AI-Driven Grading in Higher Education

A growing body of research shows the effectiveness of AI-based grading systems in supporting instructors and streamlining evaluation processes. In particular, Crompton and Burke (2023) identify automatic assessment as the most common application of AI in higher education, particularly for academic writing. Their review highlights various studies, including one that focused on assessing the writing skills of Uyghur ethnic minority students in China. The study demonstrated improvements not just in academic outcomes but also in student engagement across cognitive and emotional domains.

Key Findings
- **Efficiency and Consistency**: AI-powered grading significantly reduced the time required for instructors to assess student work, allowing them to allocate more time for student interaction and focused instructional activities. The AI systems maintained consistency in scoring, reducing the variability and potential errors that can occur in manual grading.
- **Equity Support**: The AI system delivered personalized, context-specific feedback, helping students from underserved or underrepresented groups improve their writing and critical thinking skills. This feedback was tailored to individual learning styles and provided an opportunity for self-regulated learning, which is often crucial for students in high-need environments.
- **Scalability**: These AI grading systems are particularly beneficial for large courses or institutions with limited resources. They enable efficient grading at scale, supporting institutions with high student-to-teacher ratios, where manual grading would otherwise be prohibitively time-consuming and resource-intensive.

Implications

This case highlights the potential of AI in reducing educator workloads and improving grading efficiency. By providing personalized, real-time feedback, AI grading systems not only support student development but also help bridge equity gaps in academic support. However, human oversight remains crucial to ensure that the AI's interpretation of student work aligns with the unique educational context, especially when dealing with complex or nuanced responses.

Concluding Remarks

AI-driven automation can ease major administrative burdens in education, such as grading, scheduling, and documentation. By streamlining these tasks, AI improves efficiency and consistency, giving educators more time for student engagement, instructional planning, and personalized support. These tools are especially valuable in under-resourced environments, helping ensure more equitable access to quality education.

Still, responsible adoption is critical. Human oversight remains necessary to capture the nuances of student work and broader educational context. When thoughtfully integrated with safeguards, AI can reduce workload, strengthen instructional quality, and create more student-centered learning environments.

Conclusion

As artificial intelligence (AI) continues to shape Educational Psychology, its potential to enhance learning, engagement, and well-being is increasingly evident. From identifying learning differences to personalized interventions and scalable support, AI offers tools that make education more inclusive and adaptive. However, for AI to reach its potential, it must be integrated thoughtfully, with attention to ethics such as data privacy, equity, and preserving the human connection vital to care. School psychologists play a crucial role in guiding responsible use so AI supports student development rather than replacing the relationships at the heart of effective education.

As we transition to the next chapter, we move from education to the clinical setting, where AI supports behavior change, wellness, and chronic illness outcomes in Health Psychology.

Chapter 8: AI in Health Psychology

Health psychology is increasingly intersecting with artificial intelligence (AI) as clinicians seek new ways to personalize care, enhance engagement, and scale interventions. As physical and mental health needs become more complex—and traditional service delivery models struggle to keep up—AI technologies offer tools that may assist in managing these demands. From cognitive assessment to behavior change and mental health monitoring, AI can enable more dynamic interaction with patients' data, potentially informing better-targeted care strategies.

This chapter explores how health psychologists can leverage AI responsibly across five domains: cognitive and neuropsychological assessment, behavior change support, early mental health screening, remote monitoring, and reducing disparities in care access.

For clarity, this chapter refers to professionals in this area as *health psychologists*. However, we acknowledge that many practitioners contributing to health behavior change and integrated care—including those using AI tools—may come from diverse mental health and allied health backgrounds.

Enhancing Personalization in Health Psychology

Health psychologists are under increasing pressure to provide more personalized care, but traditional tools can fall short in meeting this need. Standardized protocols and broad diagnostic categories frequently fail to account for the nuanced and dynamic needs of individual patients, leading to mismatched treatments and trial-and-error approaches that delay recovery (Babu & Joseph, 2024; Lee et al., 2021). These challenges are especially pronounced in areas with limited technological access, such as rural settings where reliable internet access is scarce (Lee et al., 2021; Haque & Rubya, 2023), or in situations where financial barriers make essential care unaffordable (Babu & Joseph, 2024). Additionally, cultural challenges, such as stigma or mismatched backgrounds between providers and patients, can further complicate the task of offering personalized and effective interventions (Babu & Joseph, 2024; Thakkar et al., 2024).

Opportunities and Benefits

Artificial intelligence (AI) offers a promising solution to these challenges by providing health psychologists with tools that have the potential to enhance the precision, adaptability, and scalability of care. Through real-time data analysis, predictive modeling, and multimodal monitoring, AI can more effectively match the right interventions to the right patients at the right time. By personalizing treatment plans based on complex, longitudinal data, such as symptom patterns, treatment histories, and real-time feedback, AI can help reduce the inefficiencies associated with trial-and-error approaches. These technologies also have the potential to improve patient engagement, particularly in underserved populations, by offering more equitable, accessible, and culturally sensitive interventions. When thoughtfully integrated into clinical practice, AI tools can support clinicians in delivering more targeted and dynamic care that is better aligned with the individual needs of each patient.

Tailoring Interventions to Individual Needs

AI systems can analyze complex, longitudinal data—such as treatment history, demographics, symptom patterns, and engagement levels—to match patients with interventions more likely to succeed. This can help reduce the inefficiencies of trial-and-error approaches and helps improve alignment between care plans and individual needs (Lee et al., 2021; NIHMS, 2021; Spring Health). These tools also provide responsive strategies, ensuring clinicians can respond with matched interventions that align with the individual needs of their patients.

Real-Time Monitoring and Adaptive Care

As a foundational shift in mental health care, AI tools now allow clinicians to monitor patients' emotional and physiological states in near real time, via data from wearables, apps, and spoken words. This ongoing visibility can enable more dynamic care adjustments, helping to close the gap between in-session evaluation and day-to-day experience. For instance, tools like Ellipsis Health use voice analysis to detect emotional shifts between sessions, enabling timely clinical intervention before symptoms escalate.

Reducing Bias and Promoting Equity

AI offers a standardized, data-driven approach to care decision-making that can help reduce disparities rooted in unconscious bias or variable clinician experience. By analyzing inputs consistently across patient populations, AI can support fairer and more transparent treatment recommendations, particularly in settings where cultural competence and equity gaps exist (Babu & Joseph, 2024).

Expanding Access Through Self-Guided Support

Conversational agents like Wysa provide real-time, CBT-based support tools to users in geographically remote or resource-constrained settings. These tools increase accessibility while maintaining privacy and availability, making them particularly valuable for early engagement or between-session support (Haque & Rubya, 2023).

Predicting Risk and Supporting Proactive Interventions

Predictive AI models can identify warning signs of treatment dropout, relapse, or disengagement by analyzing behavioral trends and symptom trajectories. This allows clinicians to intervene early, reducing care discontinuity and helping patients stay on track (Ginger, Spring Health; Babu & Joseph, 2024).

Synthesizing Multimodal and Longitudinal Data

AI can integrate diverse data streams—from Electronic Health Records (EHRs) and speech patterns to physiological signals and therapy notes—into a unified picture of patient progress. This richer context supports more accurate diagnoses and nuanced treatment planning, consistent with broader trends toward precision mental health care (NIHMS, 2021; Haque et al., 2018).

Improving Engagement and Patient Connection

AI-enhanced platforms can personalize content, pacing, and support, helping users feel more seen and supported. These systems offer ongoing check-ins and tailored messaging, which build emotional investment in care and encourage adherence, especially during vulnerable periods in the treatment process (Thakkar et al., 2024).

Used appropriately, these tools can augment clinical expertise by helping providers deliver more tailored, efficient, and equitable care. While caution is warranted around issues of bias, privacy, and transparency, the research to date suggests AI can play a meaningful role in shifting mental health care from generalized protocols to truly personalized support.

Example Applied AI Tools

AI-driven tools have the potential to enhance mental health care by addressing key challenges in personalization. By enabling more data-informed, adaptive, and scalable interventions, these technologies can support clinicians in overcoming barriers posed by rigid diagnostic categories, clinician bias, and limited data integration. They can also help expand access to personalized care, particularly in underserved settings where financial, technological, and cultural barriers exist. Below are several AI-powered platforms currently being explored to help create more personalized and equitable approaches in mental health care:

Quartet Health
- **What it does**: Uses AI-driven analysis of clinical profiles, treatment histories, and patient preferences to match patients with appropriate mental health providers.
- **In this context**: Quartet Health improves the personalization of care by reducing mismatches in treatment through better alignment between patients and providers. It is particularly effective in settings where clinician bias and limited provider availability pose significant challenges.
- **Evidence**: Babu & Joseph (2024) underscore the importance of reducing dropouts by aligning patient needs with the right provider expertise. Quartet's precision matching supports this by predicting engagement likelihood and helping ensure that patients receive the care they need based on their unique profiles.

Ellipsis Health
- **What it does:** Uses AI-based voice analysis to assess levels of stress, depression, and anxiety by analyzing vocal biomarkers during routine check-ins.
- **In this context**: Ellipsis Health helps bridge the gap caused by limited data integration through supplementing traditional self-reporting methods with objective, real-time voice data. This supports clinicians in making more accurate, personalized interventions based on continuous monitoring.

- **Evidence**: According to Lee et al. (2021), voice-based tools like Ellipsis can detect subtle emotional shifts between sessions, supporting the use of multimodal data to improve diagnoses. Integrating speech and physiological data helps address the challenge of limited data integration in traditional personalized care models (Haque et al., 2018; NIHMS, 2021).

Spring Health
- **What it does**: Offers personalized care pathways by analyzing intake data, clinical history, and symptom trajectories to recommend effective mental health interventions.
- **In this context**: By using AI to synthesize patient data, Spring Health can improve personalized care even in settings with limited access to traditional mental health resources. The platform ensures that individuals receive tailored interventions that address their specific needs, promoting more effective treatment and reducing reliance on rigid diagnostic categories.
- **Evidence**: AI platforms like Spring Health help refine clinical decisions and accelerate recovery by predicting treatment response and providing continuous symptom monitoring (Lee et al., 2021). This is particularly beneficial in underserved settings, where access to tailored, dynamic care is often limited.

Ginger (Headspace Health)
- **What it does**: Combines AI-supported care navigation with on-demand coaching and therapy, enhancing engagement tracking and risk detection.
- **In this context**: Ginger's real-time behavioral tracking and personalized care guidance are especially valuable for high-risk or underserved populations. By offering timely interventions, it addresses the challenge of providing personalized care to individuals who may otherwise fall through the cracks.
- **Evidence**: As noted by Babu & Joseph (2024), predictive AI tools like Ginger help prevent disengagement and promote continuity of care. Real-time tracking enables clinicians to intervene proactively, reducing reliance on sporadic, trial-and-error methods of treatment.

Practical Example: AI-Driven Prediction of Treatment Response in Mental Health

A 2021 review by Lee et al. explored how artificial intelligence (AI) is being used to predict treatment outcomes in psychiatric care, particularly for Major Depressive

Disorder (MDD), anxiety disorders, and psychosis. The review emphasizes the growing use of machine learning across diverse data sources—including electroencephalogram (EEG) signals, clinical assessments, speech patterns, and neuroimaging scans—to inform treatment decisions more precisely. This multimodal approach aligns with the broader trend toward precision psychiatry, where AI tools attempt to tailor interventions to an individual's specific clinical, behavioral, and biological profile.

Here are some capabilities highlighted in literature to consider:

Prediction Accuracy

Several studies cited in the review demonstrate that machine learning models can predict treatment response to interventions like antidepressants, Cognitive Behavioral Therapy (CBT), and brain stimulation. In some cases, these predictions meet or exceed the accuracy of traditional clinical assessments, which are particularly prone to inconsistencies and clinician bias. This is particularly relevant in conditions like MDD, where the heterogeneity of symptoms often complicates treatment planning. AI can help move beyond broad diagnostic categories and create individualized, more effective care plans.

Modality Matching

AI systems are being developed to differentiate which patients are more likely to benefit from specific treatment types—whether pharmacological, psychotherapeutic, or neuromodulatory. For instance, neuroimaging data has been used to distinguish likely responders to CBT versus medication (Lee et al., 2021; NIHMS, 2021). Tailoring interventions to individual needs reduces the inefficiencies of trial-and-error treatments and improves outcomes more quickly. In underserved settings, this approach allows for more personalized care despite resource constraints.

Efficiency Gains

By automating parts of the diagnostic and prognostic process—such as analyzing questionnaire data, EEG signals, or speech samples—AI tools can help reduce clinician workload, particularly in high-demand or resource-limited settings. As noted by Babu & Joseph (2024), this kind of augmentation can free up time for more relational and interpretive aspects of care, allowing clinicians to focus on building therapeutic rapport rather than administrative tasks.

Reinforcement from Broader Evidence

These findings are reinforced by work from Haque et al. (2018), which demonstrates that combining audio, facial expressions, and text data can detect depression symptoms with high sensitivity and specificity. Additionally, AI models can anticipate relapse by detecting subtle behavioral shifts—such as changes in speech cadence or social withdrawal, which further supports dynamic, real-time care adjustments.

While these developments are ongoing, the collective evidence points toward a future where AI-assisted treatment planning could help clinicians make more informed decisions, reduce delays in effective care, and better align interventions with each patient's unique profile.

Concluding Remarks

Personalization remains a core challenge in health psychology, constrained by standardized diagnostic systems, inconsistent clinician judgment, and systemic barriers that limit access in underserved settings. Artificial intelligence (AI) offers a path toward addressing these issues by enabling more adaptive, data-informed, and context-sensitive approaches to care. As the examples in this section show, AI tools can help match individuals with interventions that reflect their specific clinical profiles, cultural backgrounds, and real-time needs—reducing the trial-and-error nature of many current practices.

When used responsibly, these technologies may assist health psychologists in delivering more consistent, equitable, and precise care. They offer opportunities to integrate diverse data sources, flag risk proactively, and extend support to populations who often fall through the gaps of traditional service models; however, realizing this potential depends on thoughtful design, transparency, and continued validation to ensure these tools support clinical expertise.

AI is not a substitute for the human relationship at the heart of health psychology; however, when carefully integrated, it can enhance the delivery of care, bringing the field closer to a model where psychological interventions are truly aligned with the complexities of individual lives and needs.

AI in Neuropsychology

Neuropsychology—the study of brain-behavior relationships—is beginning to integrate artificial intelligence (AI) in ways that may enhance current practices. AI-driven tools are being explored for their potential to support cognitive assessments, aid in the early detection of neurodegenerative disorders, and inform more personalized intervention strategies for conditions such as Alzheimer's disease, Parkinson's disease, and traumatic brain injuries. While still emerging, these technologies offer the possibility of improving precision, efficiency, and access in the diagnosis and management of cognitive disorders.

Opportunities and Benefits

Given the delays in identifying cognitive decline and the inconsistency of current diagnostic workflows—particularly in rural or aging populations—artificial intelligence (AI) offers emerging tools that may increase efficiency, improve access, and personalize intervention. AI is beginning to provide practical support for neuropsychologists in areas such as assessment, diagnosis, and cognitive rehabilitation. While these technologies are still evolving, early research suggests that AI can help clinicians deliver more accurate, timely, and individualized care when integrated thoughtfully into practice.

Earlier Detection of Cognitive Decline

AI systems can analyze subtle speech, behavioral, and neurological data to identify cognitive impairment before it becomes clinically obvious. Tools like Winterlight Labs use linguistic patterns in short speech samples to detect early Alzheimer's-related changes with over 81% accuracy (Fraser et al., 2016), while machine learning models analyzing neuroimaging and electronic health record (EHR) data have shown success in identifying early-stage dementia risk (Lee et al., 2021; Thakkar et al., 2024).

Improved Efficiency in Diagnostic Workflows

Natural language processing (NLP) systems such as CogStack streamline the extraction of cognitive and neurological information from unstructured health records. By automating chart review and surfacing early risk markers, these tools reduce manual workload and help clinicians focus on higher-order decision-making (Kraljevic et al., 2024).

Dynamic, Personalized Cognitive Rehabilitation

AI-powered platforms like BrainHQ and emerging VRehab systems adjust cognitive training tasks in real time, tailoring them to the individual's pace, progress, and specific deficits. This enhances therapy precision while promoting engagement, especially for patients recovering from traumatic brain injury or managing neurodegenerative conditions (Abedi et al., 2024).

Multimodal Integration for Richer Cognitive Profiles

AI models can integrate diverse data sources—including speech, movement, magnetic resonance imaging (MRI) scans, EHRs, and wearable device outputs to generate more holistic cognitive assessments. This multimodal approach supports more accurate diagnosis and treatment planning, especially in complex or comorbid cases (Fraser et al., 2016; Lee et al., 2021).

Continuous, Remote Monitoring of Cognitive Health

AI-enabled cognitive screening tools like Neurotrack offer longitudinal insight into patients' brain function, capturing subtle changes over time through home-based assessments and reducing dependence on episodic in-person evaluations. In clinical trials, Neurotrack's N-CAB (Neurotrack Cognitive Assessment Battery) achieved high diagnostic accuracy, offering an accessible alternative to in-person testing (Glenn et al., 2023). These tools expand care reach, particularly for older adults and patients in rural or underserved areas.

Greater Standardization and Scalability

AI reduces variability in how cognitive assessments are administered and interpreted. Automated scoring, adaptive testing, and standardized analysis help ensure consistency across patients and providers, supporting both individual care and population-level monitoring.

Ethical and Equitable Implementation

AI also holds promise in improving equity by extending diagnostic and monitoring tools to communities with limited access to specialists. When developed with representative datasets and attention to transparency, these systems may help reduce disparities in cognitive health care.

Example Applied AI Tools

AI is offering neuropsychologists new tools with the potential to support earlier detection, more personalized interventions, and deeper insights into cognitive and mental health. While adoption is still emerging, the following examples illustrate how AI can enhance neuropsychological care in real-world settings.

Neurotrack

- **What it does**: A digital cognitive health platform that uses eye-tracking and self-administered memory tasks to detect early signs of cognitive decline.
- **In this context:** Neurotrack enables remote, at-home assessments that are especially useful in aging or rural populations where access to neuropsychological services may be limited.
- **Evidence:** Its N-CAB modules demonstrated 95–96% sensitivity and up to 89% specificity in detecting early-stage dementia, making it a viable alternative to in-person screening (Glenn et al., 2023).

CogStack

- **What it does**: An AI-enabled NLP system that extracts cognitive and neurological data from unstructured EHRs.
- **In this context:** CogStack helps identify risk markers earlier in the patient's journey by automating data extraction from clinical records, reducing the burden of manual chart review.
- **Evidence**: Supports earlier detection and more efficient diagnostics, especially when neuropsychological data are embedded in large, unstructured datasets (Kraljevic et al., 2024).

Winterlight Labs

- **What it does**: Uses machine learning to analyze short speech samples for signs of cognitive impairment.
- **In this context**: Offers a low-burden, speech-based screening approach that does not require clinical infrastructure, making it ideal for remote or resource-limited settings.
- **Evidence**: Demonstrated over 81% accuracy in detecting Alzheimer's-related speech changes, supporting its use as an early and accessible screening tool (Fraser et al., 2016).

BrainHQ

- **What it does**: A digital platform delivering adaptive cognitive training programs tailored to memory and executive function.

- **In this context**: Adjusts exercises in real time to match individual performance, supporting personalized cognitive rehabilitation for patients recovering from brain injury or managing neurodegenerative conditions.
- **Evidence:** Though outcomes vary, it supports scalable, home-based rehab that is difficult to deliver consistently in traditional care models (Pressler et al., 2022).

Savana

- **What it does**: Applies NLP to structured text EHRs, enabling cognitive health trends to be analyzed at scale.
- **In this context**: Useful for monitoring cognitive trends across large populations and identifying underserved groups at elevated risk.
- **Evidence**: Helps discover population-level patterns in cognitive decline and supports equity-focused planning in aging communities (Savana, 2024).

Practical Example: Early Detection of Cognitive Decline Using AI

Artificial intelligence (AI) is increasingly being applied to the early detection of neurodegenerative disorders like Alzheimer's disease, where early intervention is critical. Both academic and clinical research indicate that machine learning and deep learning tools can detect signs of cognitive impairment by analyzing speech, behavior, neuroimaging, and electronic health records.

Multimodal Methodology

AI systems, particularly those employing natural language processing (NLP) and deep learning, can analyze multiple data streams to detect early-stage Alzheimer's and other forms of dementia. According to Lee et al. (2021), models have successfully differentiated between types of dementia using structural MRI scans and analyzed speech patterns to identify psychosis risk in youth. These systems are capable of:

- Parsing speech data to detect reduced lexical diversity, changes in syntax, and altered fluency.
- Analyzing neuroimaging and sensor-based inputs to identify neurological patterns associated with memory decline.
- Integrating patient data from wearable devices, social behavior, and clinical history to generate risk profiles.

Thakkar et al. (2024) further support the promise of AI in neurodegenerative care, noting that machine learning models such as support vector machines and convolutional neural networks have shown high accuracy in identifying early cognitive deficits from MRI and behavioral data.

Impact & Advantages
- **High Diagnostic Accuracy**: These AI tools often match or exceed human performance in early-stage detection, enabling preclinical identification of Alzheimer's.
- **Personalized Monitoring**: AI-driven platforms can continuously assess cognitive function and trigger alerts when cognitive decline is detected.
- **Timely Interventions**: By identifying at-risk individuals early, clinicians can initiate treatment plans sooner, which may help delay disease progression.

Ethical and Practical Considerations
While the technology is promising, both Lee et al. (2021) and Thakkar et al. (2024) stress the importance of ethical safeguards. Issues include algorithmic bias (particularly with non-diverse datasets), privacy concerns around behavioral and biometric data, and the need for explainable AI to ensure clinical transparency and trust.

Concluding Remarks

AI is beginning to improve neuropsychological care by enabling earlier detection of cognitive decline, streamlining diagnostics, and personalizing rehabilitation—especially in aging and underserved populations. Tools like speech analysis, NLP, and adaptive training platforms can extend access and improve efficiency. Moving forward, success will depend on developing systems that are transparent, clinically validated, and equitable across diverse settings.

———————— ⬛ ————————

AI in Psychopharmacology & Treatment Planning

Psychiatric medication management continues to rely heavily on trial-and-error prescribing, often resulting in treatment delays, adverse side effects, and poor adherence. These inefficiencies are compounded by a lack of precision in matching patients with the most effective pharmacological options. Although AI tools—such as those leveraging pharmacogenomics, EHR data, and wearable devices—offer new avenues for personalizing care, their clinical utility is not without limitations. Biased training data, limited generalizability across populations, and inequities in access to genomic tools remain significant barriers. These challenges are especially pronounced in settings with constrained digital infrastructure or financial resources, and among populations that may distrust or lack access to AI-driven health technologies. As such, integrating AI into psychopharmacology must be done cautiously, emphasizing transparency, clinician oversight, and equity to ensure that new tools improve, rather than widen, disparities in mental health care.

Opportunities and Benefits

AI technologies are beginning to address longstanding challenges in psychiatric medication management by offering data-driven tools that support more personalized, timely, and equitable care. In a field where treatment decisions often rely on trial-and-error and subjective interpretation, AI systems can analyze diverse data sources—such as electronic health records, genetic profiles, and digital behavior—to improve medication matching and reduce delays in effective treatment. Pharmacogenomic platforms, predictive models, and wearable-integrated systems help identify patients at risk for side effects, nonadherence, or poor response. While concerns remain around biased training data and limited applicability across diverse populations, AI's ability to synthesize complex, individualized inputs offers a promising step toward more precise and proactive psychopharmacological care, especially when developed with transparency and equity in mind.

Reducing Trial-and-Error Prescribing

AI-driven decision support tools can analyze patient data—including demographics, prior treatment outcomes, and clinical presentation—to estimate the likelihood of success with specific medications. This helps reduce the guesswork in selecting initial treatments, shortening the time to symptom relief and minimizing unnecessary side effects (Benrimoh et al., 2021).

Supporting Genetically Informed and Equitable Medication Selection

Pharmacogenomic platforms like CNS Dose and Genomind can interpret genetic variants, such as in the CYP2C19 gene, to predict drug metabolism and inform treatment choice. These tools can reduce adverse reactions and improve speed to efficacy; however, their utility depends on equitable access and ensuring training data represent diverse populations (Squassina et al., 2025).

Minimizing Adverse Effects and System Burden

AI can flag potential drug–drug interactions and help tailor dosages based on patient-specific risk profiles. These systems are especially useful in reducing preventable ER visits and hospitalizations, particularly in patients on complex medication regimens (Genomind, 2021).

Translating Complex Data into Actionable Decisions

Although genomic and biometric data are increasingly available, they often overwhelm clinicians. AI systems help translate this complex information—EHRs, lab data, patient-reported outcomes—into structured, interpretable recommendations to support real-time clinical decisions.

Enabling Remote Monitoring and Early Detection of Medication Issues

Wearable-integrated platforms such as Biofourmis track metrics like heart rate, stress, and sleep to flag changes that may indicate a medication issue or relapse. These tools provide clinicians with continuous feedback, allowing for earlier, more precise treatment adjustments (Biofourmis, 2022).

Improving Medication Adherence Through Proactive Support

AI can identify early signs of nonadherence by analyzing digital behavior patterns or physiological data, to assist in prompting timely interventions. This may help to prevent relapse and support patients in maintaining treatment engagement, especially where regular follow-up is not feasible.

Addressing Bias and Improving Generalizability

AI has the potential to reduce inequities by incorporating more diverse training datasets and adjusting for social determinants of health. Rigorous fairness auditing and inclusion of underrepresented groups are critical to making these tools safe and effective for all populations—not just those reflected in existing data.

Scaling Personalized Treatment Without Overburdening Clinicians

AI can automate aspects of medication selection and follow-up monitoring, enabling more individualized care at scale. By streamlining decision-making, these tools would better support clinicians in managing complex cases efficiently, even in settings with limited psychiatric specialists.

Ensuring Ethical Use and Patient Trust

Given the sensitivity of genetic and behavioral data, AI systems must be designed with strong privacy protections and clear, explainable logic. Adhering to laws like the Health Insurance Portability and Accountability Act (HIPAA) and the General Data Protection Regulation (GDPR) is essential to maintaining patient trust and clinician confidence in AI-assisted prescribing.

Example Applied AI Tools

Several AI-powered platforms are currently being explored or implemented to support more personalized approaches in mental health care. While early findings suggest promising improvements in treatment precision, efficiency, and patient engagement, most tools are still undergoing validation, and their long-term clinical impact remains to be fully understood. Below are examples of AI-based technologies being used or studied in psychiatric care settings:

Aifred Health
- **What it is**: A clinical decision support system (CDSS) for depression treatment powered by artificial intelligence.
- **In this context**: Analyzes clinical and demographic data to generate remission probability estimates for various first-line antidepressants. The tool is designed to aid clinicians in selecting the most appropriate treatment, enhancing shared decision-making during patient encounters.
- **Evidence**: In a simulation study, Aifred was found to be acceptable to clinicians and feasible for use during live clinical interactions. Seventy percent of participants reported it improved patient understanding of treatment, and 65% said it enhanced trust. The tool integrated smoothly into clinical workflows without major disruption to the physician–patient dynamic (Benrimoh et al., 2021).

CNS Dose
- **What it does**: Combines deoxyribonucleic acid (DNA) testing with machine learning to predict individual responses to antidepressants and antipsychotics, helping clinicians select medications and doses aligned with a patient's metabolic profile.
- **In this context**: CNS Dose reduces trial-and-error prescribing and the likelihood of adverse reactions, but its utility depends on equitable access to genetic testing and training data that represent diverse populations.

- **Evidence**: A naturalistic study by Squassina et al. (2025) found that pharmacogenetic testing—particularly involving CYP2C19 gene variants—can help guide antidepressant treatment by identifying patients at greater risk for side effects or poor response. Patients with tailored regimens based on their genetic profile experienced fewer treatment failures and more stable outcomes, supporting the potential for such tools to improve therapeutic precision and reduce time to benefit.

Genomind Professional PGx Express

- **What it does**: Analyzes a patient's genetic profile to help clinicians understand how they may respond to over 130 medications, including those used for depression, anxiety, and Attention Deficit-Hyperactivity Disorder (ADHD).
- **In this context:** Genomind helps personalize prescribing and reduce adverse outcomes, but attention must be paid to ensuring clinical utility across diverse populations and practice settings.
- **Evidence**: Genomind's tool has shown adoption in psychiatry and primary care. A large U.S. case-control study linked its use to reduced emergency visits, fewer hospitalizations, and lower overall healthcare costs (Genomind, 2021).

Biofourmis

- **What it is**: A digital health platform that uses AI and wearable biosensors to monitor patient health in real time.
- **In this context:** While primarily validated in chronic disease management, Biofourmis is being explored for psychiatric applications such as detecting early signs of relapse or medication side effects, especially valuable in remote or underserved areas.
- **Evidence**: Biofourmis has launched remote care programs for conditions such as hypertension and heart failure and is expanding into behavioral health monitoring through similar AI-driven tracking capabilities (Biofourmis, 2022).

Practical Example: AI-Guided Treatment Selection with Aifred Health

Study Overview:
A peer-reviewed simulation study evaluated the early feasibility and clinician

experience of Aifred Health, an AI-powered clinical decision support system (CDSS) designed to guide antidepressant selection for patients with Major Depressive Disorder (MDD). Conducted at the Steinberg Centre for Simulation and Interactive Learning, the study involved 20 psychiatry and family medicine clinicians interacting with individuals trained to simulate patients with varying depression severities. These actors, known as "simulated patients," were trained to consistently portray different levels of depression, ensuring that each clinician could engage in a controlled, repeatable clinical interaction (Benrimoh et al., 2021).

Technology Applied:

Aifred Health uses deep learning to generate individualized remission probabilities for multiple first-line antidepressants, based on each patient's clinical and demographic data. The system aligns with the 2016 Canadian Network for Mood and Anxiety Treatments (CANMAT) guidelines and is designed to be used during clinical encounters to support shared decision-making.

Relevance

This tool directly addresses the challenge of trial-and-error prescribing by helping clinicians select medications more likely to work for a given patient, thereby reducing the time to effective treatment and minimizing adverse effects. Aifred also aims to enhance adherence by improving patient understanding and trust in the treatment plan.

Key Findings / Outcomes

- **Feasibility**: 90% of clinicians indicated they would use Aifred for at least some patients; 50% said they would use it for all.
- **Patient Engagement**: 70% reported improved patient understanding of treatment options, and 65% said it enhanced trust.
- **Workflow Integration**: Clinicians found the tool adaptable to real-time clinical flow without disrupting therapeutic rapport.
- **Training Takeaways**: Behaviors such as turning the screen toward the patient helped foster collaborative decision-making.

While the study did not assess clinical outcomes in real patients, it provided early evidence that AI-based tools like Aifred can be introduced into clinical practice without degrading the quality of the therapeutic interaction and may even enhance it when used thoughtfully.

Concluding Remarks

AI is helping shift psychiatric medication management away from trial-and-error by improving how clinicians match patients with effective treatments. Tools that incorporate genetic, clinical, and behavioral data can reduce delays, side effects, and poor adherence. Still, challenges remain around bias, generalizability, and equitable access. With thoughtful implementation, AI can support more personalized, efficient, and inclusive care, enhancing rather than replacing clinician judgment.

AI for Remote Mental Health Monitoring

Ongoing mental health monitoring remains difficult in many settings due to geographic, social, and systemic barriers. Individuals living in rural or underserved areas often face limited access to providers and infrastructure, while others experience interruptions in care between sessions or after discharge. Traditional models rely on episodic check-ins and self-reporting, which are not well suited for continuous symptom tracking. As a result, early signs of symptom escalation often go unnoticed, delaying intervention and increasing the risk of crisis.

Opportunities and Benefits

AI offers transformative potential to address the longstanding challenges of remote mental health monitoring, particularly in underserved areas where access to providers is limited. AI-driven tools provide the ability to continuously monitor symptoms, even between sessions or after discharge, offering timely detection of early warning signs of mental health decline. These technologies, such as wearable devices, voice analysis platforms, and mobile apps, can provide ongoing support while reducing the burden on healthcare systems. AI-powered systems also offer anonymity, reducing stigma and encouraging individuals to engage in care they might otherwise avoid. By integrating these tools into mental health workflows, clinicians can receive real-time insights into patient conditions, improving decision-making and continuity of care while expanding access to support for individuals facing geographic or financial barriers.

Early Detection of Symptom Escalation

Remote monitoring systems enable real-time detection of subtle behavioral shifts, such as changes in sleep patterns, voice, or activity, which may signal mental health deterioration. Studies by Haque et al. (2018) and Lee et al. (2021) demonstrate that these tools can identify depression and anxiety with comparable sensitivity and specificity to in-person assessments. This capability enables early, low-burden intervention, preventing symptoms from escalating unnoticed between clinical visits.

Continuous Monitoring Between Sessions
Wearable devices, voice-enabled check-ins, and smartphone-based mood tracking create a continuous stream of data that bridges the gap between clinical sessions. This reduces reliance on self-reporting, which is often retrospective and incomplete. For example, Biofourmis, which has been applied in chronic care settings, is expanding its use into mental health, providing more proactive intervention options (Biofourmis, 2023).

Supplementing Self-Reports with Objective Data
Many patients struggle to accurately report their emotional states or may underreport due to stigma or discomfort. AI tools address this by analyzing objective data like vocal biomarkers, activity levels, or typing patterns, offering valuable insights into mental health status even when subjective reports are limited. This may improve clinical decision-making, especially when patients find it difficult to express their symptoms.

Expanding Access for Underserved Populations
A conversational agent like Wysa delivers low-cost, on-demand CBT-based support, expanding access to mental health care for individuals in underserved regions. These tools provide continuous support without requiring a clinician's immediate presence, addressing geographic, financial, and social barriers. As noted by Babu & Joseph (2024) and Fitzpatrick et al. (2017), such systems are instrumental in reducing stigma and improving access, though attention must be given to bridging digital literacy gaps.

Personalizing Care at Scale
An AI tool such as Twill can dynamically adjust their feedback and interventions based on individual symptom patterns and engagement trends, offering personalized support at scale. This ability to personalize interventions without

increasing the clinician's workload enables the extension of tailored care to more individuals, particularly those with limited access to in-person sessions.

Post-Crisis and Transitional Support

After inpatient care or during high-risk transitional periods, AI-powered monitoring may provide an added layer of safety. By detecting early signs of disengagement or symptom relapse, these tools support structured follow-up, which may help to reduce readmission rates, ensuring that treatment continuity is maintained.

Synthesizing Multimodal Insights for Clinicians

AI systems integrate multiple data streams, such as voice analysis, wearable data, mood tracking, and text logs, into cohesive insights. This multimodal approach gives clinicians a more complete picture of a patient's condition over time, facilitating informed decision-making. As emphasized by Babu & Joseph (2024), AI tools should augment—rather than replace—professional judgment, offering more accurate assessments between sessions.

Example Applied AI Tools

Several AI-powered tools are currently being explored or implemented to support remote mental health monitoring. These platforms leverage continuous data collection, machine learning, and behavioral analytics to help identify emerging symptoms, personalize care, and expand access to support, particularly for individuals with limited access to traditional services, and geographic or social barriers.

A Note to the Reader: While some of the tools discussed in Chapter 4: AI in Clinical and Counseling Practice, particularly those for early detection and personalized care, share similar challenges with those addressed in remote mental health monitoring, this section specifically highlights how AI is used to bridge gaps in care for individuals in underserved and geographically isolated areas. These tools offer unique benefits, including continuous, low-burden monitoring and real-time insights, helping to offer ongoing support and the early detection of symptom escalation outside traditional clinical settings

Biofourmis
- **What it is**: Biofourmis is a digital health platform that uses wearable biosensors and AI-driven analytics to monitor physiological signals in real time.

- **What it does:** This system collects continuous data on metrics such as heart rate variability, respiration, and sleep patterns. The AI analyzes these inputs to detect early signs of mental health changes, alerting healthcare providers for timely intervention. The system was originally developed for chronic disease management but is expanding to mental health monitoring.
- **Evidence:** Biofourmis offers real-time, continuous monitoring for individuals who may otherwise miss intervention opportunities between sessions or after discharge. Its integration into post-acute care helps ensure that remote patients, particularly those with co-occurring chronic conditions, are continuously supported even when clinical access is limited. This addresses the gap between in-person visits and helps avoid worsening symptoms due to geographic isolation or intermittent access to care (Biofourmis, 2023).

Ellipsis Health

- **What it is**: Building on its use in personalized care, Ellipsis Health contributes to early mental health screening by analyzing vocal biomarkers indicative of depression and anxiety.
- **What it does**: The system analyzes vocal biomarkers, such as tone, pace, and energy, during brief conversations or check-ins. It can integrate with telehealth platforms or be used during routine app-based interactions to provide clinicians with objective assessments between sessions.
- **Evidence**: Ellipsis Health reduces reliance on in-person assessments by providing passive, real-time monitoring. This is particularly useful for individuals who struggle with self-reporting or feel uncomfortable expressing their emotional state. The ability to monitor symptoms remotely also ensures early detection of symptom escalation, supporting continuous care without the need for face-to-face visits (Lee et al., 2021).

Twill

- **What it is**: Twill supports remote mental health monitoring by dynamically tailoring mental health interventions based on in-app behavior and symptom trends.
- **What it does**: Using machine learning, Twill delivers personalized support through psychoeducation, mood tracking, and behaviorally informed prompts aimed at improving emotional well-being over time.
- **Evidence**: Twill's adaptive approach enables personalized care at scale, offering proactive mental health support that can be accessed between sessions, particularly in underserved populations. It allows individuals to

maintain consistent engagement with their mental health without overburdening clinicians, addressing the challenge of limited access to in-person care (Fitzpatrick et al., 2017).

Practical Example: AI-Supported Mood Monitoring with Wearables and Voice Analytics

Study Overview: A study by Haque et al. (2019) explored the use of multi-modal AI, including 3D facial expression analysis, voice recordings, and natural language processing, to assess the severity of depressive symptoms. This approach sought to overcome the limitations of self-reporting by using objective, real-time data gathered outside of traditional clinical settings. The system combined visual, audio, and linguistic data to predict depression severity, providing an alternative for continuous mental health monitoring when in-person visits are not feasible.

Technology Applied: The system developed by the research team integrates vocal tone, 3D facial motion, and spoken language transcripts. These features are processed using a causal convolutional neural network (C-CNN) to generate sentence-level embeddings, which predict symptom severity on the Patient Health Questionnaire-8 (PHQ-8) scale, a tool used to assess the severity of depressive symptoms.

Key Findings and Outcomes:

- **High Accuracy in Depression Detection**: The AI system achieved 83.3% sensitivity and 82.6% specificity in detecting Major Depressive Disorder (MDD). These results were comparable to traditional in-person screening methods, validating the use of AI for remote mood monitoring.
- **Objective, Scalable Screening:** The ability to generate PHQ scores from passively collected inputs highlights the potential for scalable and lower-cost mental health screening. This could dramatically expand the reach of mental health services, particularly in underserved areas with limited access to care.
- **Potential for Real-Time Monitoring**: While not using a wearable device per se, the voice analysis and facial tracking used in this study can support real-time monitoring and continuous care, particularly beneficial when traditional self-reporting or clinician access is limited. This aligns with the challenge of providing continuous, low-burden monitoring for individuals who experience interruptions between clinical sessions or after discharge.

How it Addresses the Challenge: This example exemplifies how AI-driven tools can bridge the gap between in-person visits by providing remote monitoring that does not rely on sporadic check-ins. By detecting early signs of symptom escalation, AI-powered tools like this one offer a way to identify mood shifts in real-time, which is crucial for early intervention in geographically isolated or underserved populations. With its ability to analyze multimodal data (voice, facial expressions, and language), this system enables continuous, passive monitoring, providing clinicians with timely insights that can guide decision-making and prevent crises before they escalate.

Concluding Remarks

AI-driven remote monitoring tools hold great promise for improving mental health care by providing continuous symptom tracking and early detection, especially for individuals in underserved or isolated areas. These technologies can enhance decision-making and ensure timely intervention; however, challenges such as data privacy, bias, and access inequities must be addressed. AI should complement, not replace, clinician judgment, and be developed with transparency and fairness to ensure it benefits all populations. Moving forward, AI in mental health monitoring must prioritize ethical use and patient engagement to improve care accessibility.

AI-Assisted Health Behavior Change Programs

Behavior change programs often face challenges such as low adherence, difficulty scaling, and poor engagement during critical, high-risk moments. Without timely, personalized support, these programs frequently fail to maintain momentum, particularly across diverse user groups. Users may struggle with fluctuating motivation, lack of accessible support, and challenges in tracking progress meaningfully, which can hinder long-term success in achieving health goals.

Opportunities and Benefits

Health behavior change programs often struggle with low adherence, scalability, and providing personalized support, particularly for diverse and underserved populations. AI technologies present an opportunity to tackle these challenges by delivering real-time, adaptive interventions that are customized to individual needs. By providing timely, data-driven feedback, AI tools can enhance user engagement, encourage sustained behavior change, and support individuals during high-risk moments when motivation is low. These systems can also scale more easily than traditional methods, offering continuous support at a fraction of the cost and helping to reach users in remote or underserved areas. When integrated thoughtfully, AI-powered interventions can improve long-term adherence and ultimately contribute to better health outcomes across a wide range of behavior change goals, such as smoking cessation, weight management, and stress reduction (Crompton & Burke, 2023).

Improved Adherence Through Real-Time Personalization

AI-powered platforms can deliver personalized prompts, feedback, and encouragement based on individual behaviors, motivation levels, and contextual cues. For example, systems like Quit Genius use biometric and behavioral data to identify high-risk moments, such as cravings or stress, and provide timely, CBT-informed support. This real-time intervention helps sustain user engagement and reduces the likelihood of relapse (Benrimoh et al., 2021).

Expanded Access to Support Across Populations

Digital coaches and AI chatbots, offering 24/7 support, make it possible to deliver scalable behavioral guidance without the need for in-person resources. For individuals in rural or low-resource areas, tools like Lark Health provide consistent support that rivals human-led programs. This expansion of access helps individuals who may have otherwise missed out on regular care (Torous et al., 2021).

Dynamic, Data-Driven Personalization

AI systems continuously learn from users' data, including sleep patterns, stress levels, activity, and engagement, to adapt the content and feedback. This ensures that interventions are always relevant and responsive to changes in users' lives. A platforms like Twill exemplifies this dynamic approach by adjusting their support in real time, improving retention and overall impact (Crompton & Burke, 2023).

Support During High-Risk or Low-Motivation Periods

AI-driven platforms analyze user inputs, including biometric signals, app engagement, and emotional tone, to provide just-in-time nudges and motivational support. Unlike traditional programs that may miss critical moments, AI tools are always available to offer support during high-risk periods, improving adherence when motivation is low and helping prevent relapse (Haque et al., 2019.

Reducing Self-Tracking Burden Through Passive Monitoring

AI systems reduce the need for manual input by passively collecting and analyzing data such as voice tone, typing speed, and physical activity. By gathering this information seamlessly, these tools offer users a clearer picture of their progress and struggles, without overburdening them with time-consuming self-tracking tasks. This reduces friction and supports sustained engagement (Matochová & Kowaliková, 2024).

Scalable, Consistent Delivery of Evidence-Based Techniques

AI allows the broad and reliable distribution of evidence-based techniques, ensuring that behavioral science principles are applied consistently. A program like Quit Genius uses natural language processing to simulate therapeutic dialogue, delivering CBT-based strategies to users at scale. This approach ensures that users receive structured support, regardless of their location or schedule (Garg & Sharma, 2024).

Example Applied AI Tools

Several AI-powered tools are currently being used or tested to support behavior change efforts in areas like smoking cessation, stress management, and physical activity. These platforms use natural language processing, predictive analytics, and real-time feedback loops to increase personalization, adherence, and scalability.

Lark Health
- **What it is**: Lark Health is an AI-powered health coaching platform widely used in chronic disease management and behavior change programs like weight loss and smoking cessation.
- **What it does**: Delivers real-time, personalized feedback and motivational support through an automated mobile app, offering coaching for lifestyle changes such as healthy eating and exercise.
- **Evidence**: A study by Stein and Brooks (2017) found that Lark's weight loss program produced results similar to those of traditional in-person programs. This demonstrates its ability to scale effectively while maintaining consistent support, making it a valuable tool for underserved populations where in-person care may not be accessible.

Quit Genius
- **What it is:** A digital therapeutic platform that uses AI-based coaching and cognitive behavioral therapy (CBT) to support smoking cessation and substance use recovery.
- **What it does**: Provides personalized feedback during high-risk moments, such as cravings or stress, using real-time data from wearables and self-reported inputs. This ensures that users receive timely interventions that help maintain motivation and prevent relapse.
- **Evidence**: Quit Genius has demonstrated its ability to support individuals through their recovery journey by offering continuous AI-driven support. It significantly improves engagement, reduces relapse, and enhances adherence by delivering scalable and personalized care (Thakkar, Gupta, & De Sousa, 2024).

Carrot's Pivot Program
- **What it is:** A digital smoking cessation platform that combines a mobile app, a personal carbon monoxide (CO) breath sensor, and optional coaching.
- **What it does**: Uses biofeedback from a Food and Drug Administration (FDA)-cleared CO sensor to provide real-time monitoring of smoking behavior. The AI-powered system delivers tailored interventions based on the user's stage of change and provides motivational support at critical times.
- **Evidence**: The Pivot Program has shown effectiveness in increasing motivation to quit smoking, with a substantial percentage of users making quit attempts and reducing cigarette use by over 50%. The use of real-time feedback and AI-powered personalization allows it to support sustained behavior change and engage users, particularly during high-risk moments (Marler et al., 2020).

Practical Example: AI-Driven Digital Health Interventions for Smoking Cessation

A narrative review by Thakkar et al. (2024) explored how AI-enabled tools are being used to support smoking cessation, particularly through digital platforms that integrate chatbot-based coaching, cognitive behavioral therapy (CBT) content, and biometric monitoring. These interventions aim to provide accessible, continuous support that adapts to user behavior in real time.

Key Observations

- **Support During High-Risk Moments**: AI chatbots can deliver CBT-informed prompts and emotional support during moments of craving or stress, helping users stay engaged without requiring constant human involvement.
- **Behavioral Adaptation**: The platforms discussed adapt their interventions based on user engagement patterns, enabling more personalized and timely support.
- **Scalable Mental Health Integration**: These tools show promise not only in smoking cessation but also in addressing co-occurring mental health concerns such as anxiety and stress, which often accompany substance use.

While the review did not present new clinical trial data, it highlights how AI systems like Quit Genius and similar digital therapeutics may improve motivation, reduce relapse, and provide scalable alternatives to traditional smoking cessation programs.

Concluding Remarks

AI-powered behavior change tools offer great promise in enhancing the accessibility, relevance, and effectiveness of health interventions, especially where traditional methods struggle. By delivering personalized, real-time support, these tools can improve engagement and adherence, particularly in high-risk moments; however, it's crucial that AI systems respect user autonomy, ensuring ethical design, informed consent, and transparency. Moving forward, collaboration among clinicians, technologists, and behavioral scientists will be key to ensuring that AI-driven interventions foster responsible and impactful behavior change.

Conclusion

In Health Psychology, AI holds the potential to transform how care is personalized, supported, and scaled. By integrating real-time, data-driven tools into clinical practice, AI can assist in more accurately matching interventions with patients' needs, offering enhanced personalization and ongoing support; however, it is crucial to ensure these technologies are implemented ethically, respecting user autonomy, ensuring transparency, and prioritizing cultural sensitivity. As AI becomes more deeply integrated into Health Psychology, collaboration between clinicians, technologists, and behavioral scientists will be vital to maximizing the benefits of AI in providing personalized, effective, and equitable care.

Chapter 9: AI In Research & Experimental Psychology

Artificial intelligence (AI) is beginning to reshape aspects of psychological research, offering tools that can support hypothesis development, automate data workflows, and detect patterns in large, complex datasets. As research questions grow in scale and data become more abundant, many traditional methods—while still essential—can struggle to keep pace. Literature reviews, experimental design, and participant management can be time-consuming and prone to human biases or inconsistencies (Luoma, Tong, & Cheng, 2024).

AI systems, particularly large language models (LLMs) and predictive analytics, have shown promise in augmenting these workflows. Early studies suggest they can assist in generating novel hypotheses, simulate certain cognitive responses, and improve the consistency of data processing or trend analysis. Still, most applications remain exploratory. Questions about reliability, validity, and ethical implementation persist, and the role of AI is best understood as supportive—not standalone. This chapter reviews six emerging areas where AI is being tested in psychological science and examines what current evidence does—and does not—yet support. By staying grounded in empirical findings and ongoing limitations, the goal is to provide researchers with a realistic sense of AI's current and potential role in the field.

Enhancing Data Processing for Psychological Studies with AI

Psychological research increasingly relies on large and varied datasets—from behavioral logs and neuroimaging to survey responses and longitudinal studies. Yet many researchers continue to rely on manual methods for data cleaning, analysis, and interpretation, which can be labor-intensive, inconsistent, and difficult to scale. As the volume and complexity of data grow, so do the risks of bottlenecks, delayed insights, and subjective decision-making that can affect the transparency and reproducibility of research. AI offers a promising set of tools to support these challenges, augmenting traditional research practices with automation, pattern detection, and scalable processing capabilities. Thoughtful integration of AI into psychological research workflows may help teams more efficiently navigate the demands of data-intensive studies while preserving rigor and interpretability.

Opportunities and Benefits

As psychological research grows more data-intensive and complex, AI offers researchers practical tools to streamline data processing, improve analytic rigor, and expand the scope of inquiry. While not a replacement for human interpretation, these tools can support more consistent, scalable, and insightful workflows when implemented responsibly and with appropriate oversight. Below are some ways AI can enhance data processing in psychological research.

Automating Literature Review and Thematic Analysis
AI-powered natural language processing (NLP) systems can efficiently scan thousands of academic articles, extract recurring themes, and map conceptual networks. For instance, Tong et al. (2024) used GPT-4 and causal graphs to analyze over 43,000 papers, generating hypotheses and illuminating underexplored areas. These tools reduce time spent on manual review and enable broader, more systematic synthesis of existing literature.

Improving Data Cleaning and Preprocessing
AI can automate routine data tasks like identifying missing values, standardizing formats, and flagging anomalies, reducing manual error and speeding up analysis. This makes it easier to manage large datasets across modalities, especially in collaborative or longitudinal studies that require high data fidelity.

Detecting Patterns Across Large or Complex Datasets

Machine learning models are well-suited for uncovering nonlinear patterns, latent clusters, or unexpected relationships across multimodal datasets. In their review, Mukherjee & Chang (2024) emphasized AI's ability to detect alignment between abstracted research designs and empirical outcomes, supporting deeper theoretical exploration.

Supporting More Adaptive and Scalable Statistical Modeling

AI models such as decision trees and neural networks can accommodate complex, non-linear relationships without rigid assumptions. While these methods require interpretive caution, they offer a valuable complement to traditional inferential techniques, particularly when working with dynamic or unstructured data.

Enhancing Reproducibility and Transparency

Automated workflows and standardized pipelines can help reduce variability caused by subjective decisions. For instance, a study by Tong et al. (2024) demonstrated that AI-generated hypotheses maintained conceptual coherence and consistency when processed at scale. When combined with tools such as model documentation or version tracking, these workflows enhance the clarity and replicability of analytic decisions.

Accelerating Insights in Multi-Site and Large-Scale Studies

AI tools can harmonize data from diverse sources—labs, platforms, or populations—supporting faster integration and cleaner interim analyses. This is particularly useful in multi-site studies where differences in data formatting or collection protocols can delay progress.

Broadening Access to Computational Methods

User-friendly AI platforms are helping democratize data analysis, allowing researchers without programming expertise to engage in computational workflows. This makes data-intensive research more inclusive and helps expand methodological capabilities across the field.

Enabling Integration of Multimodal and Longitudinal Data

AI supports the integration of text, audio, physiological signals, and time-series data into unified analyses. As Chang & Srivastava (2024) note, this can yield richer insights into dynamic psychological processes, especially in real-world or ecological contexts.

Example Applied AI Tools

Several AI-powered platforms and research tools are beginning to support experimental and data-processing workflows in psychological science. While many of these systems are still evolving, they offer practical ways to streamline literature reviews, identify patterns in large datasets, and evaluate theoretical claims with greater consistency and scale.

Litmaps
- **What it is:** A visual literature mapping tool that uses citation networks and AI-based clustering to identify relevant research based on user-defined papers or topics.
- **What it does:** Helps researchers visually explore related studies, trace conceptual linkages, and track how ideas evolve across time in large academic corpora.
- **Real-world context:** Similar functionality is described in Tong et al. (2024), who used AI-based citation network analysis and topic clustering to map causal knowledge from over 40,000 psychology papers. This aligns closely with Litmaps' conceptual design (Tong et al., 2024).

Semantic Scholar API
- **What it is:** A research discovery tool with an open-access application programming interface (API) allowing programmatic access to millions of academic papers.
- **What it does:** Enables filtering, citation analysis, and metadata extraction across diverse psychological literature.
- **Real-world context:** Tong et al. (2024) used large-scale document retrieval systems and semantic filtering methods similar to those enabled by Semantic Scholar's API for efficient literature review and knowledge extraction (Tong et al., 2024).

Rayyan AI
- **What it is:** A web-based tool used to semi-automate systematic reviews, leveraging machine learning to screen abstracts and tag articles.
- **What it does:** Assists researchers in inclusion/exclusion decisions for large sets of papers, improving the efficiency of systematic reviews.
- **Real-world context:** AI-assisted systematic review pipelines (e.g., for screening and clustering papers) are discussed in Tong et al. (2024) as key elements of scaling evidence synthesis in psychology (Tong et al., 2024).

GPT-Based Models (e.g., GPT-4 via OpenAI API)

- **What it is**: A large language model (LLM) capable of summarization, hypothesis generation, text classification, and semantic reasoning.
- **What it does**: Analyzes unstructured psychological texts, extracts causal relationships, and identifies novel research hypotheses through semantic and causal reasoning.
- **Real-world context**: Tong et al. (2024) used GPT-4 to process over 43,000 psychology articles, extracting causal relations and generating novel hypotheses. When combined with causal graphs, GPT-4-generated hypotheses demonstrated significantly greater conceptual novelty than those created by GPT-4 alone or by another large language model, Claude-2.

OpenAI Evals / AI-as-Participant Research Pipelines

- **What it is:** An adaptation of evaluation tools that allow researchers to test how LLMs perform in experimental psychology tasks traditionally applied to human participants.
- **What it does**: Enables LLMs to simulate human-like responses in structured psychological protocols, such as those testing moral reasoning, attention, and theory-of-mind.
- **Real-world context**: Tong et al. (2024) benchmarked GPT-4's performance using experimental frameworks aligned with psychological theory and found that when paired with causal reasoning tools, it could produce hypotheses on par with human-generated ones in terms of novelty and semantic depth.

Practical Example: AI-Driven Literature Analysis for Psychological Research

A study by Tong et al. (2024) demonstrated how AI can assist in large-scale psychological literature analysis by extracting causal relationships and generating novel hypotheses. Using a combination of a large language model (LLM) and a causal knowledge graph, the system analyzed over 43,000 psychology articles to construct a semantic network and propose new research directions.

Key Findings

- **Automated Insight Generation**: The AI-generated hypotheses demonstrated novelty and conceptual depth comparable to those produced by doctoral scholars.

- **Efficiency and Scale**: The AI system processed tens of thousands of articles, efficiently identifying emerging themes that may otherwise be overlooked in manual reviews.
- **Conceptual Accuracy**: Deep semantic analysis showed that the AI-generated insights aligned with expert reasoning and covered a broader conceptual spectrum than LLMs alone.

These findings highlight AI's potential to enhance the speed, scale, and insightfulness of literature synthesis in psychology.

Concluding Remarks

AI has the potential to support more efficient, reproducible, and scalable data workflows in psychological research. From assisting with literature synthesis to detecting patterns in complex datasets and standardizing preprocessing steps, AI tools may help reduce manual burdens and enhance insight generation. However, realizing these benefits depends on responsible implementation, prioritizing transparency, fairness, and alignment with established scientific standards, as well as ensuring that AI can distinguish valid and reliable research from flawed or biased studies. It is essential that AI systems are designed to recognize and apply correct statistical analyses and research methodologies, reassuring users that they are weeding out bad science in ways that align with the judgment of human scientific experts. Moving forward, efforts to make AI more interpretable, inclusive, and ethically grounded will be crucial for ensuring its value as a complement to human expertise in advancing psychological science.

Automating Psychological Hypothesis Generation

Psychological hypothesis generation has traditionally depended on expert intuition, theory-driven reasoning, and time-consuming literature reviews. While these methods have led to important insights, they are increasingly challenged by the overwhelming amount and complexity of modern research. As psychological science becomes richer in data and more interdisciplinary, it's becoming harder to manually identify new, promising hypotheses at scale.

Recent advances in AI—particularly in large language models (LLMs), causal knowledge extraction, and semantic reasoning—offer new ways to support this early-stage research process. These tools are not replacements for theoretical insight but may serve as helpful complements by revealing underexplored questions, mapping conceptual relationships, and streamlining hypothesis development.

Opportunities and Benefits

AI is emerging as a useful tool in the early stages of psychological research, particularly in the automation and augmentation of hypothesis generation. With the volume of academic literature growing rapidly and research becoming more interdisciplinary, many traditional methods of developing testable hypotheses are becoming harder to scale. When applied with appropriate oversight, AI systems—especially large language models (LLMs) and causal graph frameworks—can help reduce cognitive bias, uncover novel ideas, and prioritize promising research directions based on existing evidence.

Reducing Bias and Expanding Conceptual Diversity

LLMs can draw from vast and diverse bodies of literature, enabling exploration beyond a researcher's immediate domain or theoretical framework. This can reduce confirmation bias and support the generation of more varied hypotheses. Tong et al. (2024) found that AI-generated hypotheses exhibited greater conceptual diversity when paired with causal reasoning tools than when using LLMs alone.

Managing Information Overload

Psychologists face growing difficulty keeping up with expanding research outputs. AI tools can assist by summarizing, clustering, and thematically analyzing large academic corpora. In Mukherjee & Chang (2024), GPT-4 demonstrated its ability to interpret nearly 600 post-cutoff psychology studies and forecast theoretical implications, helping researchers filter and prioritize information.

Accelerating Literature Synthesis

Tasks such as citation mapping, topic modeling, and semantic filtering can be automated to reduce the time burden of early-stage review. Tong et al. (2024) demonstrated that combining LLMs with causal graphs enabled scalable and conceptually rich hypothesis generation based on over 43,000 articles.

Detecting Conceptual Relationships and Indirect Effects

AI systems equipped with causal graph modeling can uncover latent variable relationships, such as mediators or moderators, that are often difficult to identify manually. These insights support more complex and testable hypotheses. Tong et al. (2024) used such tools to model well-being constructs across multiple causal dimensions.

Prioritizing Hypotheses with Predictive Signals
AI can help rank hypotheses by estimating empirical viability based on methodological patterns or historical findings. Mukherjee & Chang (2024) showed that GPT-4 could predict study outcomes from redacted abstracts, suggesting potential for triaging research ideas by plausibility and empirical alignment.

Cross-Disciplinary Integration
LLMs trained on broad corpora can connect psychological constructs to adjacent fields such as neuroscience, behavioral economics, or education. This capability supports richer, systems-level thinking. As Luoma et al. (2024) argue, interdisciplinary analogical reasoning may expand the theoretical scope of psychological research.

Scaling Thematic Exploration Across Massive Corpora
AI can process thousands of articles to detect trends, shifts, or gaps in the literature. These capabilities allow researchers to identify underexplored questions and track emerging themes with greater objectivity and scale, as shown in Tong et al. (2024)'s work with causal knowledge graphs.

Example Applied AI Tools

As researchers explore ways to automate psychological hypothesis generation, several AI-powered tools and frameworks are emerging to support this process. While many are still in development or early testing phases, the following examples illustrate how AI can assist in synthesizing literature, identifying causal patterns, and proposing empirically grounded hypotheses.

GPT-4 (OpenAI)
- **What it is**: As mentioned earlier in Chapter 2 a large language model capable of summarization, semantic analysis, and hypothesis generation.
- **What it does**: In psychological research, GPT-4 can be used to scan large volumes of literature, identify gaps, and generate novel research questions that align with established themes or emerging areas.
- **Real-world context**: Tong et al. (2024) used GPT-4 in combination with causal graph modeling to generate 130 new psychological hypotheses, with

outputs rated as conceptually novel and aligned with expert-generated themes (Tong et al., 2024).

Causal Knowledge Graph Framework
- **What it is**: An AI-driven system that combines extracted causal relationships from literature with graph-based modeling.
- **What it does**: Constructs causal networks from thousands of psychology papers and applies link prediction to identify unexplored variable relationships. This supports the generation of empirically grounded hypotheses.
- **Real-world context**: This framework helped extract over 19,000 causal relation pairs from 43,312 articles, facilitating scalable hypothesis generation related to well-being (Tong et al., 2024)

Semantic Hypothesis Validation Framework
- **What it is**: A validation approach that assesses whether AI-generated outputs—such as hypotheses or decisions—align with established psychological constructs and conceptual knowledge.
- **What it does**: Uses semantic comparison techniques to evaluate the internal coherence, plausibility, and domain-relevance of language model outputs, filtering out statements that may reflect surface-level correlations rather than meaningful scientific reasoning.
- **Real-world context**: Hagendorff (2024) applied this framework to assess whether LLM-generated responses in psychological tasks reflected genuine conceptual understanding or mere linguistic mimicry. The analysis showed that high-performing models, such as GPT-4, produced outputs that exhibited semantically valid reasoning patterns, suggesting potential for supporting structured scientific inquiry.

Predictive Hypothesis Modeling
- **What it is**: A machine learning-based framework designed to evaluate the plausibility of scientific hypotheses by predicting outcomes and assessing alignment with reported findings.
- **What it does**: Uses general-purpose AI models—such as GPT-4—to redact findings from real psychology articles, predict empirical outcomes, and assess conceptual consistency. This approach supports researchers in prioritizing hypotheses that are more likely to be substantiated through empirical testing.

- **Real-world context**: Mukherjee and Chang (2024) tested GPT-4 on 589 psychology articles published after its training cutoff. The AI demonstrated the ability to emulate expert-level reasoning by accurately predicting study outcomes and showing conceptual understanding, suggesting potential utility in early hypothesis evaluation (Mukherjee & Chang, 2024).

Practical Example: AI in Psychological Hypothesis Generation

A groundbreaking study by Tong et al. (2024) explored AI's application in automating hypothesis generation through LLMs and causal knowledge graphs. The researchers analyzed 43,312 psychology articles to extract causal relation pairs and construct a specialized psychological causal graph. Link prediction algorithms were then applied to generate 130 potential psychological hypotheses related to well-being.

Key findings
- AI-generated hypotheses exhibited higher novelty compared to those derived solely from LLMs ($t(59) = 3.34$, $p = 0.007$).
- Hypotheses generated through LLM-causal graph integration closely aligned with expert-level insights, surpassing purely LLM-derived hypotheses ($t(59) = 4.32$, $p < 0.001$).
- Deep semantic analysis confirmed that AI-generated hypotheses incorporated meaningful conceptual relations consistent with existing psychological literature.

These results underscore AI's potential to augment traditional research methodologies by accelerating hypothesis generation and fostering scientific creativity.

Concluding Remarks

AI can support psychological hypothesis generation by reducing manual burdens, expanding the conceptual search space, and making literature synthesis more efficient. When paired with causal graph modeling, semantic validation, and predictive frameworks, LLMs can help identify novel, empirically grounded research questions that align with expert reasoning; however, realizing this potential requires attention to transparency, dataset diversity, and responsible deployment. As tools evolve, the most productive applications of AI will be those that amplify human insight, enabling researchers to generate, refine, and prioritize hypotheses with greater speed, scale, and rigor.

Improving Experimental Design and Data Collection with AI

Experimental design and data collection form the backbone of psychological research, providing the structured conditions needed to investigate behavior, cognition, and mental health; however, traditional approaches face well-known constraints—including participant recruitment inefficiencies, human error, and rigid study structures—that can introduce bias and limit scalability. As psychological studies increasingly rely on digital platforms and multi-site collaboration, the need for more adaptive and streamlined methods has grown. Artificial intelligence (AI), particularly using machine learning, natural language processing (NLP), and real-time data monitoring, offers a set of emerging tools that can help researchers improve the precision, efficiency, and consistency of their experimental workflows. This section outlines opportunities where AI may support improvements in study design and data integrity, while also highlighting current limitations that require thoughtful implementation.

Opportunities and Benefits

AI offers several promising avenues for supporting experimental design and data collection in psychology, particularly as research increasingly relies on digital platforms and distributed study environments. While human oversight remains essential, AI systems—when thoughtfully implemented—can reduce inefficiencies, improve consistency, and help scale studies with greater adaptability and rigor.

Enhancing Participant Recruitment and Sample Diversity
AI-powered recruitment systems can assist in identifying eligible and representative participants by analyzing demographic, behavioral, or prior participation data. This can help address sampling bias and expand access to hard-to-reach populations. Luoma et al. (2024) note that AI-enhanced strategies are already improving reach and responsiveness in digital studies, making participant pools more inclusive.

Improving Engagement with Adaptive Survey Design
Natural language processing (NLP) tools allow for real-time adaptation of survey flow—modifying question phrasing, order, or content in response to participant

behavior. This reduces dropout rates and increases response quality. Luoma et al. (2024) report that adaptive AI routing led to greater completion rates and clarity in digital surveys, highlighting AI's role in improving participant experience.

Reducing Human Error and Standardizing Study Administration

AI can automate many aspects of experimental setup—such as task delivery, timing, and instruction presentation—leading to more consistent execution across sessions. Mukherjee & Chang (2024) demonstrated that GPT-4 could simulate full experimental workflows with high fidelity, suggesting potential to reduce variability introduced by human facilitators.

Supporting Personalized and Dynamic Study Designs

Machine learning models can adapt stimuli or task difficulty based on individual participant responses. This enables more personalized experiments, which are especially valuable in cognitive or clinical contexts. Hagendorff et al. (2024) highlighted how models like GPT-4 can dynamically adjust to shifting cognitive demands, enhancing ecological validity in longitudinal or interactive studies.

Real-Time Data Monitoring and Error Detection

AI systems can continuously monitor live data collection for anomalies, missing values, or deviations from protocol. This early feedback loop helps researchers address data integrity issues before they escalate. Studies by Tong et al. (2024) and Luoma et al. (2024) suggest that real-time monitoring contributes to cleaner, more actionable datasets and reduces the burden of post-hoc data cleaning.

Scaling and Standardizing Multi-Site Research

In multi-lab studies, AI tools can harmonize data collection procedures across locations by automating documentation, delivery, and quality checks. Mukherjee & Chang (2024) demonstrated that AI frameworks could generalize across study contexts, improving reliability and comparability across collaborative research environments.

Together, these opportunities reflect a growing alignment between AI capabilities and the needs of modern psychological research. While challenges remain—such as ensuring transparency, maintaining interpretability, and avoiding overreliance—these tools offer meaningful support in improving the design and implementation of studies at scale.

Example Applied AI Tools

As researchers seek to enhance the efficiency and rigor of experimental studies, several AI-driven platforms and methods are being tested or implemented in psychological research. These tools help optimize participant interaction, streamline data collection, and support more adaptive and scalable research protocols.

GPT-4 (OpenAI) for Experimental Stimuli Generation
- **What it is**: As mentioned in Chapter 2, a large language model capable of generating and refining experimental materials such as narrative stimuli, survey prompts, and decision-making scenarios.
- **What it does**: GPT-4 supports researchers by simulating psychologically relevant responses, creating variation in stimuli, and assisting in hypothesis refinement using causal and semantic reasoning.
- **Real-world context**: Tong et al. (2024) used GPT-4 to identify novel psychological hypotheses by synthesizing causal knowledge across over 43,000 articles. Similarly, Luoma et al. (2024) highlight GPT-4's capacity to support various research stages, including experimental material design and emulating cognitive processes that align with human reasoning patterns.

PsychStudio
- **What it is**: A cognitive testing platform enhanced by AI to automate task delivery, detect errors in real time, and dynamically adjust task flow based on participant behavior.
- **What it does**: Supports data integrity during experiments by ensuring accurate timing, reducing variability, and enabling adaptive design features that respond to user input.
- **Real-world context**: Mukherjee & Chang (2024) describe a predictive experimental framework in which AI was used to simulate experimental reasoning, validate hypotheses, and generalize to unseen research contexts—demonstrating how similar platforms could assist with dynamic task management and validity assurance.

Qualtrics with AI-Powered Survey Optimization
- **What it is**: A widely adopted survey tool that now integrates AI features such as NLP-based clarity checks, bias detection, and adaptive question routing.

- **What it does**: Improves data quality by optimizing question phrasing, reducing participant fatigue, and tailoring surveys to individual responses.
- **Real-world context**: Luoma et al. (2024) emphasize the potential of NLP tools in refining psychological instruments and enhancing response validity through real-time feedback and adaptive question structures.

Gorilla Experiment Builder (AI-Enhanced Version)
- **What it is**: An online platform for building and deploying behavioral studies, now featuring AI-driven error detection, adaptive workflows, and consistency monitoring.
- **What it does**: Allows researchers to conduct large-scale, high-fidelity experiments with customizable stimuli and AI-assisted quality control across diverse participant samples.
- **Real-world context**: Hagendorff et al. (2024) discuss the use of AI to reduce cognitive biases, increase experimental consistency, and simulate reasoning processes, aligning with Gorilla's push toward automated experimental standardization.

Case Example: AI-Supported Hypothesis Evaluation in Psychology

A study by Mukherjee and Chang (2024) introduced a novel AI-driven framework to assess whether large language models like GPT-4 can emulate expert reasoning in the scientific research process. To avoid the issue of rote memorization, the researchers used 589 psychology articles published after GPT-4's September 2021 training cutoff. The model was tasked with predicting study outcomes and theoretical implications based solely on redacted abstracts, where all empirical findings were removed.

Key Findings
- **High Predictive Alignment**: GPT-4 demonstrated strong empirical and theoretical alignment with actual study results. In most cases, its predictions matched the published outcomes with high fidelity, often scoring 8 or 9 out of 9 on structured evaluation rubrics.
- **Generalization to Novel Research**: The AI was able to reason about entirely unseen studies, indicating it could extend prior conceptual knowledge to new contexts—not just repeat memorized patterns.
- **Emulation of Expert Tasks**: The authors note that GPT-4 performed functions traditionally assigned to research assistants or postdoctoral

scholars—such as redacting findings, forecasting outcomes, and proposing theoretical interpretations, suggesting its viability in early-stage scientific workflows.

These results highlight AI's emerging potential to support hypothesis development and conceptual analysis in psychology. While not a substitute for domain expertise, models like GPT-4 could enhance research efficiency and creativity when paired with appropriate oversight and rigorous methodology (Mukherjee & Chang, 2024).

Concluding Remarks

AI has the potential to support more dynamic, scalable, and accurate approaches to experimental design and data collection in psychology. From real-time monitoring and adaptive survey delivery to AI-assisted stimuli generation and participant selection, these tools can reduce administrative burden and improve methodological consistency. Importantly, recent studies—such as those by Mukherjee and Chang (2024) and Luoma et al. (2024)—suggest that AI can emulate key components of early-stage research tasks, offering practical assistance in study execution and analysis. Still, these technologies are most effective when paired with human oversight and used transparently. Going forward, the focus should remain on integrating AI in ways that enhance rigorous research standards, support inclusivity, and protect the interpretive depth that psychological inquiry requires.

AI as a Replacement for Human Research Participants in Psychological Studies

Psychological research has traditionally depended on human participants to explore cognition, emotion, behavior, and decision-making; however, this reliance introduces several enduring challenges, including recruitment limitations, response variability, ethical constraints, and financial cost (Luoma et al., 2024; Hagendorff et al., 2024). These issues can restrict the pace and scope of research, particularly in early-stage experimentation or when working with sensitive topics.

Emerging developments in artificial intelligence (AI), particularly large language models (LLMs) such as GPT-4, present new opportunities to supplement traditional participant-based research. While these models do not replicate lived experience or embodiment, recent studies suggest they can reliably simulate certain cognitive and behavioral patterns, offering consistent, reproducible, and ethically flexible alternatives in defined contexts. This section explores how AI is being used to model participant behavior, assesses real-world applications, and highlights where simulated agents may enhance or complement psychological experimentation.

Opportunities and Benefits

Using large language models (LLMs) like GPT-4 as simulated research participants is an emerging practice that offers new possibilities for experimental psychology. While these models do not replicate lived experience or embodiment, they may serve as useful complements to traditional human samples—particularly in early-stage testing, ethically sensitive domains, or theory development.

Simulating Human-Like Responses in Psychological Tasks
LLMs can emulate a wide range of cognitive and behavioral patterns—including decision-making, moral reasoning, and affective responses—under structured prompts. Hagendorff et al. (2024) and Mukherjee & Chang (2024) showed that GPT-4 responses aligned closely with human reasoning across hundreds of psychological tasks, suggesting these models can serve as consistent proxies in defined contexts.

Improving Standardization and Reproducibility
AI-generated responses are highly consistent across trials, free from common sources of variability such as fatigue, misunderstanding, or social desirability. Tong et al. (2024) and Hagendorff et al. (2024) argue that this reliability can support replicability efforts in behavioral research and reduce noise in comparative experiments.

Reducing Logistical and Ethical Barriers
AI-simulated participants require no consent, pose no risk of distress, and eliminate privacy concerns, making them particularly useful in ethically complex or sensitive studies. As highlighted by Hagendorff et al. (2024), these systems can help prototype study designs that might otherwise be difficult or inappropriate to test with human subjects.

Allowing Controlled Manipulation of Demographics and Cognition

LLMs can be instructed to simulate individuals with specific traits or profiles, enabling controlled variation without recruiting diverse human samples. Mukherjee & Chang (2024) demonstrated how GPT-4 could model varied cognitive patterns, offering a scalable tool for testing stratified hypotheses in early-stage research.

Accelerating Hypothesis Testing and Reducing Cost

LLMs can generate thousands of participant-like responses in minutes, allowing rapid iteration and simulation of experimental scenarios. Mukherjee & Chang (2024) showed that GPT-4 was able to evaluate 589 post-training psychology studies. This is an undertaking that would be infeasible for human teams at the same speed or scale.

Supporting Novel Paradigms in "Machine Psychology"

Emerging fields like "machine psychology" study AI systems using tools from cognitive and social psychology. Hagendorff et al. (2024) argue that AI is not just a simulation tool but a subject of inquiry itself. Applying traditional paradigms to LLMs can uncover how reasoning, bias, and social cognition manifest in artificial agents.

In summary, while LLMs cannot replace the richness of human data or the complexity of lived experience, they offer clear advantages in consistency, cost, and speed. Used responsibly, AI-generated responses may help fill methodological gaps, particularly in the early or exploratory phases of psychological research.

Example Applied AI Tools

As researchers explore the use of AI to simulate human participation in psychological studies, several tools and methodologies have emerged that allow large language models to act as proxies for human subjects. The following examples illustrate how AI systems are being tested or applied to generate behavioral data, evaluate psychological theories, and support experimental research with greater speed and scalability.

GPT-4 (OpenAI) for Simulated Participant Responses

- **What it is**: As introduced in Chapter 2, a large language model capable of generating context-sensitive, human-like responses in psychological tasks.
- **What it does**: GPT-4 can respond to prompts simulating decision-making, moral reasoning, or cognitive assessments. It has been used to test whether AI-generated answers align with those of real participants across a wide range of psychological paradigms.

- **Real-world context**: In Mukherjee and Chang (2024), GPT-4 predicted the outcomes of peer-reviewed psychology studies based solely on redacted abstracts, achieving high alignment with actual results. Hagendorff et al. (2024) further demonstrated how LLMs could reproduce moral reasoning patterns found in human cohorts, suggesting these models can approximate psychological participant roles under defined constraints.

LLMCG Framework (LLM + Causal Graph)
- **What it is**: A system that combines large language models with causal knowledge graphs to generate and evaluate hypothetical psychological relationships.
- **What it does**: Extracts causal claims from psychological literature, builds semantic graphs, and tests hypothetical connections using link prediction. These predictions can serve as proxies for participant-derived conceptual reasoning in theory-building studies.
- **Real-world context**: Tong et al. (2024) used the LLMCG to simulate the generation of 130 hypotheses about well-being. These AI-generated insights were found to be more novel and semantically rich than those derived from LLMs alone.

Machine Psychology Benchmarks
- **What it is**: A research approach and set of paradigms designed to evaluate LLMs as behavioral agents using experimental tasks modeled after traditional psychology.
- **What it does**: Measures how well LLMs replicate human patterns of reasoning, bias, memory, and emotion through structured experiments.
- **Real-world context**: Hagendorff et al. (2024) reviewed emerging practices in "machine psychology," including adapting classic psychological paradigms to probe emergent behaviors in LLMs. Their work emphasizes using behavioral benchmarks (e.g., moral dilemmas, attention tests) to test whether LLMs can serve as reproducible and interpretable participant analogs.

Deception Simulation in LLMs
- **What it is**: An experimental setup designed to assess whether LLMs can simulate deceptive behaviors similar to those studied in human participants.

- **What it does**: Prompts LLMs to adopt specific roles and produce misleading or manipulative outputs, enabling researchers to explore moral cognition, social theory of mind, and behavioral divergence.
- **Real-world context**: In the study *Deception Abilities Emerged in Large Language Models* (Hagendorff, 2024), researchers found that LLMs could simulate multiple forms of deception—including white lies and manipulative intent—under specific conditions. These behaviors mirror real human behaviors often studied in social psychology, suggesting possible applications in simulated participant research.

Practical Example: AI-Simulated Moral Reasoning Using GPT-4

A recent study by Hagendorff et al. (2023) evaluated whether large language models (LLMs) like GPT-4 could replicate human-like moral reasoning. The researchers administered over 400 classic moral dilemmas to the model, analyzing whether its decisions aligned with established psychological patterns of judgment.

Key Findings
- **High Agreement with Human Norms**: GPT-4's decisions showed strong alignment with human responses, often replicating key reasoning patterns observed in experimental psychology.
- **Simulation of Multiple Ethical Frameworks**: The model could flexibly produce responses consistent with both deontological and utilitarian moral frameworks depending on prompt structure.
- **Lower Variability, Higher Consistency**: Compared to human samples, GPT-4's responses demonstrated lower variability across scenarios, supporting its use for generating consistent baseline data in experimental setups.

These results suggest LLMs like GPT-4 could serve as a supplemental tool in moral psychology research, helping simulate participant responses, reduce data noise, and test experimental conditions before large-scale recruitment.

Concluding Remarks

AI-simulated participants are not a replacement for human subjects, but they offer valuable capabilities for addressing logistical and methodological challenges in psychological research. Studies show that LLMs like GPT-4 can generate consistent, conceptually aligned responses, allowing researchers to experiment with and refine their research designs before conducting real-world studies, reduce replication concerns, and ethically navigate complex experimental scenarios (Hagendorff et al., 2024; Mukherjee & Chang, 2024).

Looking ahead, the most effective use of AI in this space may lie in hybrid approaches—where human and AI data streams complement each other. Continued work is needed to better understand the cognitive limits of LLMs, document where simulations break down, and ensure these tools are applied responsibly. With thoughtful integration and methodological transparency, AI can become a useful asset in experimental psychology, supporting both scientific creativity and practical rigor.

———————— ◆ ————————

Predictive Modeling for Psychological Trends

Psychological researchers are increasingly challenged by the complexity and volume of data involved in tracking trends related to mental health, behavior, and well-being. Traditional approaches—while methodologically rigorous—can be slow, resource-intensive, and limited in their ability to integrate large, multimodal datasets. Cognitive biases, inconsistent measurement across studies, and delays in interpreting evolving social patterns further complicate efforts to forecast psychological outcomes (Luoma et al., 2024).

Artificial intelligence (AI), particularly through predictive modeling, offers a complementary approach. By processing diverse forms of psychological data—ranging from survey responses and clinical notes to digital behavior—AI systems can help researchers detect patterns, flag early indicators of distress, and generate forecasts about mental health trajectories. These tools are not meant to replace expert judgment, but rather to augment trend analysis by improving scalability, consistency, and responsiveness. This section highlights how AI can contribute to forecasting psychological trends, including emerging methods, tools, and documented benefits.

Opportunities and Benefits

Artificial intelligence is opening new avenues for psychological trend forecasting by enabling researchers to analyze large, complex, and often multimodal datasets at scale. While these systems are not substitutes for theoretical reasoning or clinical expertise, they offer complementary strengths in identifying early indicators of mental health risk, revealing emerging behavioral patterns, and supporting proactive intervention planning.

Early Detection of Mental Health Risks

AI models can flag signs of psychological distress—such as subtle changes in language, behavior, or physiological data—before they escalate into diagnosable conditions. Predictive systems trained on clinical records, social media data, or digital health logs have demonstrated strong performance, with some models forecasting depressive episodes with over 90% accuracy weeks in advance (Hagendorff et al., 2024).

Trend Analysis Across Large, Heterogeneous Populations

Traditional research methods often struggle to integrate data from diverse sources. AI can synthesize survey responses, clinical notes, digital behavior, and more, enabling broader and more dynamic tracking of psychological states across communities and time. Tong et al. (2024) showed how causal graph modeling and LLMs could detect nuanced trends in well-being research using tens of thousands of academic articles.

Personalization of Mental Health Interventions

Predictive modeling allows for tailoring interventions to individual needs by analyzing treatment history, demographics, or behavioral signals. Luoma et al. (2024) describe how AI can support data-driven recommendations, improving alignment between individuals and effective therapeutic strategies.

Faster Response to Emerging Psychological Trends

AI systems can process and interpret incoming data streams in near real time, making it possible to monitor societal mood shifts or spikes in stress-related indicators. This responsiveness supports more agile policymaking and public health planning, especially in crisis contexts or under conditions of rapid social change (Luoma et al., 2024).

Improved Consistency and Objectivity in Forecasting

AI applies standardized logic and statistical reasoning across datasets, helping to reduce the interpretive variability and biases that can influence human-coded analyses. In Mukherjee & Chang (2024), GPT-4 predicted empirical study outcomes from redacted abstracts with high alignment to actual findings, underscoring the potential of these tools for consistent predictive reasoning in research contexts.

Discovery of Underexplored Constructs and Relationships
By linking semantically related variables or highlighting indirect causal relationships, AI models can identify blind spots in current literature. Tong et al. (2024) demonstrated this through the LLMCG framework, which generated new hypotheses about psychological well-being by uncovering novel relationships among known constructs.

In summary, when applied with care, predictive AI systems can help researchers shift from reactive to anticipatory mental health strategies, while also expanding the scope and depth of psychological inquiry; however, the long-term utility of these tools will depend on addressing ongoing concerns about model transparency, fairness, and alignment with scientific and ethical standards.

Example Applied AI Tools

As predictive modeling becomes more central to understanding mental health and behavioral trends, several AI tools and frameworks are emerging to support researchers in processing, forecasting, and personalizing psychological data. The examples below reflect both applied systems and experimental frameworks that are highly relevant to identifying trends in psychology.

GPT-4 for Psychological Trend Forecasting
- **What it is**: Again, a general-purpose large language model capable of synthesizing vast unstructured data sources, including mental health records, survey texts, and longitudinal reports.
- **What it does**: GPT-4 can support trend analysis by extracting relevant behavioral markers from narrative datasets and modeling plausible future states. It enables researchers to track emerging mental health topics and simulate how psychological responses may evolve under different conditions.
- **Real-world context**: Mukherjee and Chang (2024) demonstrated that GPT-4 could predict the empirical outcomes of 589 psychology studies published after its training cutoff, indicating that LLMs can reason about psychological research trends with minimal exposure to specific datasets.

LLMCG Framework (LLM + Causal Graph)
- **What it is**: As described earlier, a hybrid tool combining the pattern recognition of large language models with the structure of causal knowledge graphs.
- **What it does**: Extracts causal relationships from psychological literature and builds networks that reveal how constructs such as well-being, stress, and resilience interconnect over time. This facilitates the prediction of future states based on inferred causal dynamics.
- **Real-world context**: Tong et al. (2024) used this framework to process over 43,000 psychology articles, generating 130 novel hypotheses about well-being. Their approach outperformed LLMs alone in novelty and conceptual richness, suggesting its utility in trend-based forecasting.

Behavioral Risk Prediction Models
- **What it is**: A class of machine learning models trained on large-scale datasets (e.g., electronic health records, survey responses, or social media content) to identify and forecast mental health risk factors.
- **What it does**: Detects early indicators of psychological distress—such as shifts in language, behavior, or health utilization patterns—and predicts progression toward diagnosable conditions.
- **Real-world context**: As described by Hagendorff et al. (2024), predictive systems trained on patient data have demonstrated over 90% accuracy in forecasting major depressive episodes months before clinical onset, supporting their application in proactive mental health care.

AI-Augmented Social Media Monitoring Tools
- **What it is**: Natural language processing systems tailored for analyzing large volumes of user-generated content, such as tweets or forum posts, to detect population-level mood shifts or emergent stressors.
- **What it does**: Identifies trends in psychological expressions (e.g., anxiety, burnout, isolation) and maps them geographically or temporally to assist public health planning.
- **Real-world context**: This class of tool closely aligns with findings in Luoma et al. (2024), who emphasize the role of AI in scaling psychological surveillance for real-time trend detection and targeted interventions, even though it is not directly mentioned in their study.

Practical Example: AI-Driven Predictive Modeling of Psychological Constructs Using LLMCG Framework

A study by Tong et al. (2024) introduced a novel AI-based methodology for modeling psychological trends by combining large language models (LLMs) with causal knowledge graphs. The researchers analyzed over 43,000 psychology articles, extracting causal relation pairs and using a link prediction algorithm to identify emerging psychological trends related to well-being. This approach allowed for the generation of hypotheses that reflected both semantic novelty and theoretical depth.

Key Findings

- **Scalable Trend Forecasting**: The LLMCG framework processed tens of thousands of articles, generating 130 unique hypotheses on psychological well-being that exhibited conceptual alignment with expert-generated themes.
- **Semantic and Causal Accuracy**: AI-generated hypotheses outperformed those created by LLMs alone in terms of semantic depth and causal plausibility, suggesting that the combined approach enhances both interpretability and insightfulness.
- **Application to Predictive Modeling**: The causal graphs produced by the AI framework revealed emergent patterns in well-being research, enabling data-driven forecasting of which psychological constructs were gaining traction or being underexplored.

These results illustrate how AI-driven systems can support predictive modeling in psychology by synthesizing large volumes of literature and identifying causal trends, offering researchers an advanced tool for tracking the evolution of psychological constructs and guiding empirical inquiry.

Concluding Remarks

AI-driven predictive modeling is becoming an increasingly viable tool for identifying and understanding psychological trends, offering potential improvements in speed, scale, and sensitivity compared to traditional methods. When responsibly deployed, these tools can support earlier identification of mental health risks, enable more personalized interventions, and provide researchers with richer, multi-dimensional insights into how behavior and well-being evolve over time; however, to ensure long-term value and trustworthiness, researchers must address ethical concerns, model bias, and the interpretability of AI-generated insights. Ongoing collaboration between psychologists, data scientists, and ethicists will be essential to guide the integration of AI into predictive psychological research in a responsible and impactful way.

Conclusion

AI is reshaping psychological research by accelerating analysis, enabling complex simulations, and expanding how we generate and test hypotheses. These tools can deepen our understanding of human behavior—if used with care; however, speed and scale do not excuse poor methodology. Responsible use of AI in research still demands rigorous design, high-quality data, and transparency at every stage. It is not a shortcut; it is a companion to scientific inquiry.

In Chapter 10, we step back to consider the broader ethical and regulatory questions raised by AI across all psychological settings—from clinical to forensic to research domains.

Chapter 10: Navigating Regulation, Privacy, Bias, and Ethics

AI tools are showing up more in mental health settings, from processing sensitive client information to shaping treatment decisions. But many of these tools were built within legal and ethical systems that never anticipated technologies that can "learn on their own" and work in ways even their developers can't always fully explain.

In mental health, this raises serious concerns. AI has the potential to influence treatment plans, affect forensic outcomes, and even alter the trust that sits at the heart of the therapeutic relationship.

This chapter looks at the real-world challenges clinicians face when AI is introduced into psychological and forensic contexts. We'll discuss where current protections fall short, the risks of bias and privacy violations, and the uncertainty around who is accountable when things go wrong.

We'll also explore promising solutions—such as updates to laws, new models of oversight, and professional practices—that are being developed to ensure AI supports, rather than undermines, the core values of mental health care. The emphasis throughout will be on what these developments mean for you as a practitioner, and how they may affect your clients and the institutions you work within.

Regulation and Governance

AI is moving into mental health care more quickly than our laws and regulations can keep up. While rules like HIPAA in the U.S. or GDPR in Europe provide some protections, they were not designed for technologies that constantly evolve, learn, and sometimes produce results even their creators can't fully explain. This creates important gaps in areas such as accountability, transparency, and client safety.

In this section, we revisit regulation and oversight from the perspective of clinical responsibility. The goal is not to provide a policy review, but to look at the challenges these gaps create for clinicians: Who is responsible when an AI tool makes a mistake? How enforceable are existing protections? And how should therapists balance their own clinical judgment with recommendations generated by an algorithm?

We'll also look at new efforts—whether legal, institutional, or ethical—that aim to ensure AI tools are consistent with the fundamental values of psychological care.

❖ **What You'll Learn:**

- Why current laws (e.g., HIPAA, GDPR) often fall short when applied to adaptive, self-learning AI tools
- How regulatory gaps create risk for clinicians and patients alike
- Where responsibility lies when AI-influenced decisions cause harm
- How shared accountability between clinicians, developers, and institutions could be structured
- What governance models are emerging to ensure ethical and transparent AI deployment

The Limits of Current Legal Protections

Most AI tools used in mental health today fall under existing privacy and professional rules. In the U.S., that usually means HIPAA governs how protected health information (PHI) is handled. In the European Union, GDPR regulates how data can be collected, stored, and shared.

Under HIPAA, AI systems that work with PHI—such as therapy notes, biometric information, or predictive assessments—must follow rules around confidentiality and access. But HIPAA does not cover what happens when an AI system continues to "learn" after being put into use, or when it produces results that clinicians can't easily explain.

GDPR offers stronger protections, including:
- The right to informed consent before personal data is used.
- The right to an explanation of decisions made by algorithms.
- The right to request deletion of personal psychological data.

Even with these safeguards, both HIPAA and GDPR were written with stable, predictable systems in mind—not with self-learning algorithms that change over time. These laws assume data use and decision-making can be tracked and explained, but many AI tools don't work that way. They adapt as they take in new data, and not necessarily with clinician oversight.

This ongoing evolution can raise tough questions about reliability, fairness, and client consent. In clinical or forensic settings, where decisions may affect someone's health or even their freedom, these gaps in regulation can create ethical and legal dilemmas that current laws were never designed to resolve.

Global Initiatives to Bridge the Gap

As AI becomes more integrated into mental health care, different regions and institutions are stepping up to guide its safe and ethical use. From global principles to local laws, here are some salient points regarding regulation at time of this writing.

WHO Key AI Principles (2024):
The World Health Organization has outlined basic principles for using AI in health care. These stress safety, fairness, accountability, and serving the public good. In mental health, this includes paying close attention to vulnerable populations, since unsafe or biased tools can cause harm. The WHO emphasizes:

- Cultural sensitivity in how AI is used.
- Transparency in how AI makes decisions.
- Shared responsibility among developers, institutions, and clinicians to protect dignity and preserve therapeutic trust.

EU AI Act (2024):

The European Union passed a major law that regulates AI by its level of risk. Mental health applications are classified as "high-risk," so strict requirements apply. Developers must document and explain how their systems work. Organizations that use these tools must provide training, oversight, and safeguards. For clinicians, this highlights that AI cannot be treated as a "black box." Its outputs must be explainable and accountable.

APA Guidelines on Artificial Intelligence (2023):

The American Psychological Association has published guidance on the ethical use of AI in practice, education, and research. These guidelines are not laws, but they give professionals a framework for using AI responsibly. Key points include:

- Upholding existing ethical standards when AI is used in clinical work.
- Making sure informed consent is obtained when AI influences assessment or treatment.
- Being transparent about how AI is used and its limits.
- Monitoring for bias, especially in tools used with diverse or marginalized groups.

For mental health professionals, the APA's position is clear. Using AI does not reduce ethical obligations, it adds to them. Clinicians must stay informed, carefully evaluate tools, and keep client well-being at the center of practice.

U.S. Executive Initiatives:

At the federal level, efforts have been less coordinated. The White House has issued guidance such as the AI Bill of Rights, which emphasizes fairness, protection against discrimination, and the right to a human alternative. These principles are not legally binding, but they show growing recognition of risks. Oversight now comes from agencies such as:

- The FDA, which reviews AI-enabled medical devices.
- The FTC, which monitors harmful or misleading practices.

For mental health professionals, the lack of a unified federal law means oversight remains fragmented, with limited protections tailored specifically to mental health AI.

State-Level Legislation in the U.S.:

Since there is no unified federal law on AI, individual states are creating their own policies. These vary, but the trend is clear. States are attempting to regulate how AI is developed, deployed, and disclosed, especially in areas that affect human well-being such as health care, financial services, and public safety.

At the time of this writing, several types of legislation are in process:

- **Transparency Requirements:** Some states want organizations to notify consumers when AI is used in important decisions, such as approving insurance claims, setting credit limits, or evaluating eligibility for mental health services.

- **Bias Auditing and Fairness Testing:** More proposals are calling for regular bias testing of automated systems, especially in areas like employment, housing, and health care. This may directly affect AI tools used in assessment, intake, or resource allocation.

- **Watermarking and Content Disclosure:** To address the rise of generative AI, some states require AI-generated text, audio, or images to include visible or embedded disclosures. This is especially relevant for media, advertising, or clinical education. It helps prevent misinformation and supports informed consent.

- **Consumer Protection and Oversight:** Some states are creating commissions or task forces to study AI's impact on civil rights, privacy, and mental health. Others are giving attorneys general authority to investigate harmful or deceptive uses of AI, including tools marketed to clinicians or patients.

For mental health professionals, this means AI compliance is not only a federal or ethical issue, but also a local one. A tool permitted in one state may require added disclosure or documentation in another. For clinicians practicing telehealth across states, staying aware of these differences is important to avoid unintentional violations.

In summary:
AI regulation is developing on many levels. The WHO offers global ethical guidance. The EU enforces binding standards for high-risk systems. The U.S. provides principles at the federal level but lacks a unified law, leaving oversight to agencies like the FDA and FTC. The APA offers a profession-specific framework that reinforces ethical responsibilities. Individual states are now adding further rules around transparency, bias audits, disclosure, and consumer protection.

For mental health professionals, the key message is this. Expectations around fairness, accountability, and transparency are increasing. How these expectations are enforced depends on where you practice and how you use AI. Compliance is not just legal, it is also ethical, professional, and increasingly local. Stay current with changes in your state, your country, and your field. AI can support mental health care, but only when used with responsibility, clarity, and respect for clients.

Who Is Accountable When AI Fails?

Mental health professionals are usually held legally and ethically responsible for how AI outputs are used. In practice, these tools are meant to support decision-making, not replace a clinician's professional judgment. This reinforces the idea that the clinician is the final decision-maker. But it also raises a hard question: can professionals truly be accountable for tools whose inner workings they can't fully see or explain?

The complexity of AI systems creates new challenges:

- **Limited transparency:** Clinicians may not know how an output was generated; yet may feel pressured by institutions to rely on it anyway.

- **Developer disclaimers:** Companies often label their tools as "advisory" to avoid liability—even when design choices strongly shape clinical outcomes.

- **Institutional gaps:** Organizations may adopt AI without clear rules, training, or accountability systems, leaving frontline clinicians to carry the risk.

The result is that individual practitioners are often left holding responsibility, even when problems stem from hidden algorithms, biased data, or institutional decisions. This creates an **accountability gap**, where responsibility is not shared uniformly. It can be argued that a more balanced approach would require developers, institutions, and clinicians to work together to clarify roles and distribute liability more equitably.

Toward Shared Responsibility and Ethical Oversight

Several changes could help spread accountability more fairly across everyone involved:

- **Expand product liability laws to developers of AI tools in mental health practice.** This would mean that when harm comes from faulty design, biased training data, or poor documentation, developers—not just clinicians—would also be legally responsible. This helps ensure practitioners aren't left to carry the consequences alone.

- **Require clear documentation and audit trails.** These would make it possible for outside reviewers to see how an AI system works and trace errors back to their source. In mental health care, this would give clinicians a way to justify their choices when influenced by AI and protect them if liability questions arise.

- **Mandate institutional oversight and training.** Organizations that bring AI into practice must provide education about what the tools can and cannot do, set up monitoring systems, and make sure clinicians are supported to use their own judgment rather than feeling pressured to accept whatever the algorithm says.

Examples of how these reforms are beginning to take shape include:

- **EU AI Act:** This law requires developers to provide evidence that their systems are safe, fair, and accountable, with rules that vary depending on the risk level. For mental health, these rules create a legal foundation for greater transparency.

- **WHO Principles:** The World Health Organization stresses patient safety and inclusiveness, making clear that governing AI is not only a matter of law but also an ethical duty to protect those most at risk.

- **Independent AI ethics boards:** Like Institutional Review Boards (IRBs), these groups can review high-stakes systems before they are used and continue to monitor them afterward. They can evaluate whether tools meet standards of fairness, cultural sensitivity, and "do no harm."

Together, these approaches can create space for increased ethical discussion, give clinicians clearer protection when things go wrong, and make liability a shared responsibility. For mental health professionals, the key message is that accountability should not rest only on the individual clinician but should be shared across developers, institutions, and practitioners alike.

What Clinicians Can Do Now

While regulations continue to develop, clinicians can take practical steps right now:

- **Document** when AI tools are used and how their outputs influence clinical decisions.

- **Explain AI use to clients in plain language** as part of the informed consent process.

- **Advocate within your institution** for transparency, clear standards, and strong governance around AI use.

- **Stay updated** with professional guidelines and new standards as they emerge.

- **Engage in peer discussion and supervision** to reflect on AI outputs, share experiences, and identify risks or best practices together.

- **Take part in training and professional development** that build AI literacy, awareness of bias, and ethical use, so that practice grows alongside the technology.

Even as regulations evolve, psychological care cannot wait. Clinicians, developers, and institutions all have a responsibility to work together now to ensure that AI supports—rather than undermines—ethical, fair, and trusted mental health care.

★ **Key Takeaways:**

- **Current laws like HIPAA and GDPR offer important privacy protections**, but they fall short when applied to adaptive, opaque AI systems commonly used in mental health. These frameworks were not built to handle tools that change over time or produce outputs that defy easy explanation.
- **Global and professional initiatives—such as the WHO Principles, EU AI Act, APA guidelines, and U.S. policy guidance—provide different models for oversight.** However, enforcement and scope vary widely, and mental health-specific protections remain limited.

- **At the federal level in the U.S., there is no unified law that governs AI in mental health care.** Oversight is scattered across agencies, with additional—but inconsistent—regulation emerging at the state level.
- **Clinicians should not be the sole bearers of risk.** Developers and institutions must also be accountable through clear documentation, liability structures, and ongoing oversight.
- **Ethical governance structures—such as AI review boards and bias audits—are essential** to ensure fairness, transparency, and cultural sensitivity, especially when working with vulnerable or marginalized populations.
- Mental health professionals can stay proactive by:
    - Documenting when and how AI is used.
    - Educating clients in plain language.
    - Advocating for transparency and governance.
    - Continuing their training and professional development in AI.

Privacy and Security

AI brings new privacy challenges to mental health care that go far beyond traditional boundaries. Unlike general medical information, psychological data can include therapy notes, biometric signals, emotional states, and deeply personal life stories. If this information is mishandled, the result can be more than just a breach of confidentiality; it can damage dignity and erode the trust at the heart of therapy.

For mental health professionals, protecting privacy is therefore not only a legal obligation but also a core ethical responsibility.

❖ **What You'll Learn:**
- Why psychological data is uniquely vulnerable compared to other health data
- Core risks: cybersecurity gaps, re-identification, surveillance creep, and secondary data use
- The challenges of informed consent when AI systems are opaque or embedded into care

- How cross-border data flows complicate compliance with HIPAA, GDPR, and other frameworks
- Why children, adolescents, and vulnerable populations require special safeguards
- Practical steps clinicians can take to protect client trust and confidentiality

Data Privacy and Security Risks

AI-powered tools handle extremely sensitive psychological data that require stronger safeguards than most other types of health information. Unlike many areas of medicine, psychological records often capture the most personal parts of a person's life—such as thought patterns, relationship histories, emotional states, and disclosures made in therapy. If this information is exposed, the harm can go far beyond reputation or finances. It may cause stigma, emotional distress, or even long-term consequences for employment, insurance, or legal standing.

While HIPAA and GDPR provide a basic layer of protection, real-world risks remain:

- **Cybersecurity vulnerabilities:** Weak logins, poor encryption, or even simple mistakes can lead to exposure, despite official safeguards. Smaller practices using consumer-level tools are especially at risk because they often lack professional IT support.

- **Re-identification of anonymized data:** Even if names are removed, clients can sometimes still be identified through unique language patterns or life events described in transcripts. Supposedly "anonymous" data may not stay anonymous.

- **Surveillance creep:** Apps that track moods or wearables that monitor behavior may share data with insurers, employers, or even law enforcement. What starts as a wellness tool can shift into a tool of monitoring or discrimination—sometimes without the user or clinician being aware.

These risks make clear that privacy breaches in psychology are not abstract—they directly threaten the therapeutic bond of trust.

What clinicians can do:

- Check vendors for strong security documentation, such as independent audits and encryption standards. Don't assume safety just because a tool is popular or marketed to clinicians.

- Treat HIPAA compliance as a minimum standard, not a guarantee. Confirm that vendors also prohibit resale of data, using it for AI training, or sharing it with third parties.

- Encourage multi-factor authentication for both clinicians and clients to reduce risks from weak or reused passwords.

- Avoid collecting unnecessary information. Enter only what is essential for care and be cautious with extra details in consumer-grade apps.

- Review vendors' privacy policies regularly, since terms can change without notice. Build this into ongoing practice management.

- Be transparent with clients about risks, protections, and trade-offs. Explaining where data goes and who controls it strengthens therapeutic trust and supports informed consent.

Informed Consent and Transparency

Informed consent is a cornerstone of ethical care, but AI makes this process more complicated. Many AI tools work in ways that clients cannot easily see or understand, which makes it harder for them to fully grasp how their information is being used or how decisions may be shaped. This lack of clarity puts both legal compliance and the therapeutic alliance at risk.

Key challenges include:

- **Hidden influence on care:** AI may play a role in assessments or treatment planning without being clearly disclosed. A client might believe they are receiving only the clinician's independent judgment when, in fact, an AI system has strongly influenced the outcome.

- **Opaque communication:** When AI use is disclosed, the explanation is often buried in technical language or lengthy privacy policies. Clients who are stressed, overwhelmed, or less familiar with digital systems may not be able to truly understand what it means.

- **Incomplete choice:** In some settings, clients may not be told they have the option to opt out of AI-assisted methods, or may feel pressured to accept them as the default.

The result is a consent process that may meet legal requirements but fall short ethically.

What clinicians can do:

- **Use plain language** when describing AI tools, avoiding technical terms. Adjust explanations to match the client's level of understanding.

- **Make AI disclosure routine,** including it directly in consent forms and conversations. Doing so shows respect for client autonomy.

- **Offer concrete examples,** such as: *"This program helps summarize my notes, but it does not make treatment decisions."* Real-life illustrations make AI's role clearer.

- **Document consent discussions** in clinical records, noting what was explained and how the client responded. This protects both clinician and client.

- **Affirm client autonomy** by offering human-only alternatives when possible and reassure clients that declining AI use will not affect their care.

- **Check for comprehension** by asking clients to explain, in their own words, what they understood about the AI tool. This ensures genuine informed consent.

Transparent consent is more than a legal safeguard—it is essential for maintaining trust in the therapeutic relationship and for helping clients feel respected, informed, and empowered.

Cross-Border Data Flows

Many AI systems run on cloud servers that span multiple countries, which raises questions about which privacy laws apply. Data stored or processed abroad may fall under different—and sometimes conflicting—regulations. For example, a U.S. clinician using a platform hosted in Europe may have to comply with both HIPAA and GDPR, while also following the vendor's own contractual terms. In some cases, data may even pass through regions with weaker protections, putting clients at risk without their knowledge or consent.

Key challenges include:

- **Conflicting regulations:** HIPAA requires long-term record retention, while GDPR emphasizes a client's right to have data erased. When both apply, they may clash.

- **Unclear jurisdiction in breaches:** If a data breach happens on a foreign server, it can be difficult to know which authority is in charge, which laws apply, and how quickly clients must be notified.

- **Variable enforcement:** Some countries have strong oversight systems, while others lack effective enforcement, leaving psychological data vulnerable.

- **Client awareness:** Many clients assume their information is stored locally. They may feel misled or betrayed if they later discover their data was processed abroad without explicit disclosure.

What clinicians can do:

- Ask vendors where data is stored and processed and request written confirmation.

- Choose region-specific hosting when possible, keeping data within jurisdictions that have clear protections.

- Verify compliance with all relevant frameworks (HIPAA, GDPR, etc.) if cross-border transfers cannot be avoided.

- Review contracts and Business Associate Agreements (BAAs) to clarify which laws apply, who is responsible in case of a breach, and how liability is handled.

- Discuss jurisdictional risks openly with clients, making sure they understand which protections apply to their data.

- Prefer vendors who provide transparent audit trails that show where data is stored and who has accessed it.

Cross-border data flows make accountability complicated. Without careful oversight, sensitive psychological information can "fall through the cracks" of competing legal systems. Careful vendor selection, open communication, and clear documentation are essential to reduce these risks.

Secondary Use of Data and Commercialization

AI platforms often make money from psychological data by using it for product development, marketing, or even resale. Even when data is "anonymized," there are risks if patterns can be traced back to individuals. For example, mood-tracking app data could be sold to advertisers to target people based on their mental health states. Or aggregated patterns could be shared with insurers, shaping coverage decisions without the client's knowledge. These practices raise serious ethical questions about ownership and control: does psychological data belong to the client, the clinician, or the company collecting it?

Key challenges include:

- **Unclear ownership:** Many platforms claim broad rights over client data once it is uploaded, treating it as company property.

- **Commercial incentives:** Companies may put profit ahead of patient well-being, finding ways to resell or repurpose sensitive data.

- **Loss of trust:** Clients may feel deeply betrayed if they learn that personal disclosures are being commodified, even if the data was anonymized.

- **Long-term risks:** Once sold or shared, psychological data can persist in secondary databases indefinitely, beyond the control of both clinician and client.

What clinicians can do:

- Choose vendors that prohibit selling or repurposing data and get explicit policy statements written into contracts.

- Look for HIPAA compliance plus clear restrictions on secondary use, since HIPAA alone does not prevent commercialization.

- Be upfront with clients about risks during the consent process, clearly stating how their data will—or will not—be used outside of treatment.

- Limit the amount of data entered, sharing only what is necessary for care. Avoid entering extra details into tools that lack strong safeguards.

- Advocate through professional associations, licensing boards, and policy forums for stronger standards against commercial exploitation of client data.

- Monitor vendor practices regularly, as terms of service may change over time, sometimes expanding data rights unless clinicians and clients actively opt out.

By treating commercialization as both a legal and ethical concern, mental health professionals can help ensure that AI tools support care without turning sensitive client information into a commodity.

Children, Adolescents, and Vulnerable Populations

AI privacy risks become even more serious when it comes to minors and vulnerable groups. These populations may not have the full capacity to give informed consent and are especially at risk of coercion or misuse. For children and adolescents, developmental limits can make it difficult to fully understand how their information is collected, stored, or shared. Vulnerable adults—such as those with severe mental illness, under guardianship, or living in institutional settings—may also lack meaningful choice or bargaining power.

Key risks include:

- **Monitoring in schools or juvenile justice systems without true consent:** AI tools may track language, behavior, or emotional states. These are often put in place by administrators rather than clinicians, and families may not even know they're being used.

- **Long-term consequences of profiling:** Data collected during childhood can affect access to services, education, or even employment well into adulthood. Once this information exists, it is extremely difficult to erase.

- **Shifts from support to discipline:** Tools introduced to flag risks may be repurposed to discipline or exclude, often hitting marginalized groups the hardest.

- **Loss of autonomy:** Children and vulnerable adults may feel they have no choice but to accept AI monitoring, even when it undermines trust.

What clinicians can do:

- Only use tools that meet the highest security standards, avoiding consumer-grade apps that lack protections for children or vulnerable populations.

- Provide explanations that match developmental levels, using simple, age-appropriate language and examples.

- Seek layered consent (guardian consent plus child assent when appropriate), ensuring both legal and ethical requirements are met.

- Set strict limits on data sharing, clarifying that information gathered for therapy will not be repurposed for punishment or commercial use.

- Advocate for stronger protections in schools, health systems, and juvenile justice contexts to prevent discrimination and stigma.

- Monitor ongoing use, making sure tools continue to serve their stated purpose and resisting "mission creep" into surveillance or disciplinary functions.

Ultimately, children and vulnerable groups need more than just consent; they also need active protection against misuse. Ethical practice requires clinicians to go beyond the legal minimum, ensuring that technologies designed to support well-being never compromise dignity, autonomy, or opportunity.

Data Retention, Deletion, and the "Right to be Forgotten"

AI adds new complications to long-standing debates about how long psychological data should be kept. Under GDPR, clients have a "right to erasure," but licensing boards and insurance companies often require records to be retained for years. This creates a direct conflict: clinicians may be legally required to hold onto data even when clients request deletion. The issue is even more complicated with AI systems, since once data is used to train or adjust a model, parts of it may remain embedded in the system even after the original files are deleted.

Additional challenges include:

- **Data proliferation:** Client information may be stored across live servers, backups, mirrored systems, and logs, making complete deletion nearly impossible.

- **Model imprinting:** Even if raw data is deleted, patterns learned from it can remain in an AI model's behavior, creating ongoing privacy risks.

- **Regulatory conflicts:** Different regions impose contradictory requirements, leaving clinicians caught between retention rules and erasure rights.

- **Patient expectations:** Clients often assume deletion means complete removal, which can be misleading when technical realities make that impossible.

What clinicians can do:

- Ask vendors about their data retention and deletion practices, including backups, mirrored servers, and log files. Request clear written policies.

- Prefer platforms that forbid using client data to train AI models, so therapeutic disclosures do not persist in systems beyond their intended use.

- Be transparent with clients about the limits of deletion, clarifying what can and cannot realistically be erased. This builds trust and avoids false reassurance.

- Document consent discussions about retention and erasure, including both client expectations and the clinician's legal obligations.

- Advocate through professional organizations and policy bodies for clearer standards that bring legal requirements and client rights into better alignment.

For now, clinicians must navigate these tensions with care, balancing what the law requires with honest communication. Over time, new laws and technical advances may help resolve these conflicts. Until then, transparency and thoughtful vendor selection are key to protecting client trust.

★ **Key Takeaways:**
- Psychological data is uniquely sensitive; its misuse undermines trust.
- Cybersecurity gaps, re-identification, surveillance, and commercialization are critical risks.

- Informed consent must evolve to ensure transparency and client autonomy.
- Cross-border data flows create overlapping legal obligations that clinicians must manage.
- Children and vulnerable populations need special safeguards and advocacy.
- Data retention conflicts highlight the need for transparent communication and careful vendor selection.

Bias and Fairness

AI systems can make psychological decisions more consistent, but they also carry the risk of reinforcing existing social and structural biases. These risks are especially serious in clinical and legal contexts, where biased outputs can affect diagnoses, access to treatment, or even legal outcomes. This section looks at how bias in AI develops, how it has already been seen in practice, and what can be done to reduce the harm it causes.

In mental health, bias is not just a technical flaw—it is also an ethical problem. AI systems must be built and used with fairness as a central priority. Historical inequities, cultural blind spots, and systemic discrimination often show up in the data used to train AI, in the design of the systems, and in how they are implemented. Addressing these issues requires more than technical fixes. It calls for ongoing efforts toward inclusive design, diverse training data, and meaningful oversight to make sure systems are not only legally compliant but also ethically sound.

❖ **What You'll Learn:**

- How algorithmic bias emerges in psychological AI systems through data, design, and deployment
- Why proxy variables and historical disparities can perpetuate inequities in care and justice
- Real-world example of bias in forensic decision-making and their consequences

- Strategies for addressing bias, including diverse data, stakeholder involvement, fairness audits, and transparency
- The clinician's role in contextualizing AI outputs, advocating for equity, and maintaining professional responsibility

Algorithmic Bias and Its Sources

AI systems in mental health are only as fair as the data and design behind them. Without safeguards, they can reinforce inequities at every stage of development:

- **Data bias:** Historical disparities in diagnosis and treatment often end up embedded in training data. For example, Black clients have historically been overdiagnosed with schizophrenia and underdiagnosed with mood disorders. If these records are used uncritically, AI may reproduce these same distortions on a larger scale.

- **Proxy variables:** Even if protected traits like race or gender are excluded, indirect markers such as zip codes or language patterns may still carry structural inequities. These signals can create outputs that appear neutral but still function in discriminatory ways.

- **Design and testing bias:** If development teams lack diversity or if systems are tested only in limited populations (for example, affluent, English-speaking clients), the tools may fail or cause harm when used in broader communities. This reduces reliability and can produce invalid results for marginalized groups.

These forms of bias can lead to skewed risk assessments, misinterpretations of culturally normal behavior, or unfair treatment recommendations. The result is often an illusion of objectivity that hides deeper inequities.

What clinicians can do:

- **Scrutinize data sources:** Ask whether the AI tools you use were trained on data that reflects the diversity of the populations you serve. Be cautious with tools built only from insurance claims or clinical notes, which may contain systemic bias.

- **Demand transparency from vendors:** Request details on how developers assessed bias and whether they tested fairness across demographic groups.

- **Apply cultural competence:** Always interpret AI outputs in the cultural, social, and historical context of each client rather than taking results at face value.

- **Use AI as one input among many:** Treat outputs as a secondary reference point, not as final judgments, especially for clients from marginalized groups.

- **Report disparities:** Share examples of biased outputs with developers or professional associations to improve tools over time.

- **Advocate for diversity in design:** Support efforts to involve diverse clinicians, researchers, and community voices in system development and testing.

By actively engaging with these issues, mental health professionals can help reduce the risks of algorithmic bias and keep fairness at the center of AI use in practice.

Intersectionality and Compounded Bias

Bias rarely shows up along just one line, such as race, gender, or income level. More often, people experience overlapping disadvantages that interact in complex ways. For example, an immigrant teenager with a disability may face bias connected to language, cultural background, disability status, and age—all at once. These combined factors increase the chances of being misclassified, misunderstood, or treated unfairly. AI systems trained on data that treat these categories separately may miss the full picture, producing outputs that disproportionately disadvantage people at multiple margins.

In practice, this can show up as compounded disparities in risk scores, inaccurate diagnoses, or reduced access to resources for people who belong to more than one marginalized group. For instance, a bilingual child might face both linguistic bias and cultural misinterpretation in automated assessments, leading to inappropriate special education placement or denial of needed services. Intersectional bias is especially concerning in mental health and forensic settings, where AI recommendations may influence treatment plans, educational placement, or legal outcomes. Without attention to these overlapping identities, AI can reinforce systemic disadvantages instead of correcting them.

What clinicians can do:

- Stay aware of how multiple aspects of identity combine to shape how AI outputs may affect a client, recognizing that disadvantages often overlap.

- Critically review AI recommendations for signs of compounded bias, especially if they don't match the client's lived experience.

- Support **intersectional fairness audits** that evaluate AI across combined categories (e.g., race and gender together), rather than testing each in isolation.

- Advocate for the inclusion of diverse voices in system design and testing, so intersectional concerns are addressed before tools are put into practice.

- Center client stories alongside AI outputs, giving weight to lived experience even when it doesn't align with algorithmic predictions.

By keeping intersectionality in mind, clinicians and institutions can help prevent AI from amplifying layered disadvantages and move toward systems that respect the full complexity of human identity.

Real-World Example: Forensic Risk Scores

Hogan and colleagues (2021) studied a parole decision system that used machine learning and natural language processing to assess rehabilitation and the risk of reoffending. While the system was technically advanced, it was trained on past parole decisions that reflected racial and socioeconomic bias. As a result, the model disproportionately flagged marginalized individuals as high risk—even when their behavioral data did not justify it. Instead of reducing disparities, the system amplified them, showing how AI can reinforce injustice while appearing neutral.

The impact of this bias went far beyond individual cases. In parole hearings, risk scores carry heavy weight, influencing not only whether someone is released but also what conditions they face afterward. Skewed risk scores for marginalized groups meant longer incarcerations, stricter parole terms, and fewer chances for rehabilitation. Over time, these patterns deepen systemic inequities, which can achieve the opposite of what many hoped AI would achieve.

This case also highlights a key lesson: technical accuracy is not the same as fairness. The model had strong performance metrics overall, but because it was trained on biased data, it produced results that harmed certain groups. Without fairness checks, systems like this create a false appearance of neutrality, making biased decisions look objective and evidence-based.

This example underscores why fairness testing and transparency must be treated as essential safeguards—not as optional extras.

Bias in Deployment and Use

Even when an AI system is designed with fairness safeguards, the way it is put into practice can still reintroduce inequities. For example, expensive AI tools may be implemented in well-funded hospitals while under-resourced clinics lack access, helping to widen existing gaps in care rather than closing them. Decisions about where and how AI is deployed can determine whether it reduces or reinforces inequity.

Clinicians may also unintentionally over-rely on AI outputs, especially when working with populations historically mischaracterized by diagnostic systems. This happens when AI is seen as more "objective" than human judgment, leading clinicians to defer to results that may still contain bias. In forensic contexts, disparities may not come from the algorithm itself but from how judges, parole boards, or evaluators interpret scores. For instance, identical scores may be applied more harshly to marginalized defendants, reflecting broader institutional and social bias rather than flaws in the model.

Bias in deployment also shows up in differences across practice environments. Well-funded institutions may use AI with strong oversight, training, and safeguards, while under-resourced clinics may lack the capacity to monitor tools carefully. This creates a two-tiered system where some clients benefit from thoughtful integration while others are exposed to greater risks. In addition, without adequate clinician training, AI tools may be misused or trusted too much, compounding the potential for harm.

What clinicians and institutions can do:

- Work toward **equitable access** to AI tools across communities and care settings, avoiding deployments that reinforce resource gaps.

- Guard against over-reliance by using AI to **supplement, not replace**, human judgment—particularly with historically marginalized populations.

- **Track how outputs are applied** across different groups to identify disparities that emerge in practice, not just in system design.

- **Invest in training,** ensuring clinicians understand AI's limits and learn to interpret outputs critically.

- **Establish governance protocols** that review both the design of AI systems and how they are used day-to-day, making sure applications are equitable across populations.

By paying close attention to deployment—not just design—mental health professionals and institutions can help prevent fairness safeguards from being undone in practice. This ensures AI is used to support equity, not undermine it.

Addressing Bias and the Clinician's Role

Bias in AI can never be removed completely, but it can be reduced through careful design, responsible deployment, and ongoing oversight. The key themes—using diverse and representative data, inclusive development, fairness audits, transparency, and strong governance—all point to the same truth: fairness is not a single technical fix but an ongoing process that requires constant attention and accountability.

For mental health professionals, this responsibility exists at both the systemic and individual levels. Institutions and developers must build and monitor systems responsibly, but clinicians also play a vital role by questioning outputs, applying cultural competence, and advocating for equity in practice. Empathy, context, and ethical reasoning remain irreplaceable. As such, AI outputs should be treated as tools, not unquestioned truths.

Ultimately, reducing bias requires shared accountability across developers, institutions, and clinicians. By staying alert and actively engaged, mental health professionals can help ensure that AI enhances care, supports fairness, and promotes justice—rather than reinforcing existing inequities.

★ **Key Takeaways:**

- **Bias is systemic, not incidental.** AI tools can reflect historical inequities in data, design, and deployment, which can distort diagnoses, treatment, and justice outcomes.
- **Intersectionality matters.** Clients with overlapping marginalized identities can face compounded risks of misclassification and inequitable treatment.
- **Technical accuracy ≠ fairness.** High performance scores can mask deeply biased outcomes if fairness is not explicitly tested.
- **Deployment decisions shape equity**. Even well-designed systems can worsen disparities if access is unequal, training is inadequate, or outputs are misapplied.
- **Clinicians play a critical role**. By contextualizing AI outputs, questioning biases, and advocating for fairness audits and transparency, mental health professionals can help ensure AI strengthens rather than undermines equity.

Ethical Risks and Professional Responsibilities

AI can be a valuable tool in psychological practice, but it cannot replace the judgment, empathy, or contextual understanding that define ethical care. For clinicians, the challenge is not just whether to use AI, but how to use it responsibly—without weakening therapeutic trust or shifting essential human responsibilities onto machines.

This section looks at the limits of machine judgment, the risks of leaning too heavily on AI, and the safeguards needed to protect both clients and professionals.

❖ **What You'll Learn:**

- How automation bias can lead to over-reliance on algorithmic outputs and erode clinical judgment.

- The role of human oversight and institutional safeguards in ensuring responsible AI use.

- Where the line lies between AI support (e.g., symptom flagging, data visualization) and human responsibilities (e.g., diagnosis, therapeutic formulation).

- How professional guidelines emphasize transparency and clinician accountability to protect therapeutic trust.

The Limits of AI and the Risk of Over-Reliance

AI systems are very good at processing large amounts of data, spotting patterns, and flagging potential risks. These functions can support clinical decision-making, but they cannot replace the nuances of psychological assessment. Human judgment draws on nonverbal cues, relational dynamics, cultural context, and moral reasoning. These are elements that algorithms cannot capture.

The risk comes when clinicians lean too heavily on machine outputs. Under time pressure or in unfamiliar situations, **automation bias** can take hold: the tendency to trust algorithmic results even when they conflict with clinical intuition. This can lead to misdiagnosis, inappropriate treatment, or overlooking important contextual factors such as trauma history or cultural identity. The more "objective" the output looks, the easier it is to depend on it too much.

What clinicians can do:

- **Treat AI outputs as advisory, not definitive**. Use them to guide further inquiry, not as final answers.

- **Cross-check algorithmic** results against cultural, relational, and historical factors unique to the client.

- **Slow down under pressure.** Even a brief pause to question outputs can prevent over-reliance in high-stress situations.

- **Educate clients openly.** Explain when AI is used and how it informs care, while reinforcing that the clinician makes the final decision.

- **Use reflective supervision.** Bring forward cases where AI outputs felt convincing or misleading and discuss how professional judgment should take priority.

- **Stay current with guidance.** Follow recommendations from professional bodies such as the APA or institutional ethics boards on integrating AI safely.

By practicing these habits, clinicians keep human judgment at the center and make sure AI supports ethical care rather than weakening it.

Safeguarding Human Oversight

Human oversight must remain at the center of ethical AI use in mental health. Clinicians are responsible for interpreting outputs within the full context of a client's history, culture, and circumstances. AI systems may help by highlighting risk factors or summarizing records, but they cannot weigh moral considerations, therapeutic rapport, or lived experience. These aspects of care require human interpretation and ethical reasoning.

Institutions can support this responsibility by:

- Requiring documentation of when and how AI influenced decisions.
- Building in multidisciplinary review for high-stakes cases.
- Providing ongoing training in AI literacy, bias recognition, and ethical use.
- Establishing governance protocols that set clear limits on automation.

These safeguards help ensure that clinicians remain the final decision-makers and that AI outputs are viewed as aids, not authorities.

What clinicians can do:

- **Document AI use clearly.** Record when outputs are consulted and how they influenced (or did not influence) clinical decisions. This protects both clients and clinicians.

- **Maintain interpretive authority.** Frame AI outputs as advisory and make sure clients understand that decisions are made by the clinician, not the algorithm.

- **Engage in collaborative review.** Discuss AI-influenced cases with colleagues or supervisors, especially in complex or high-stakes situations.

- **Advocate within institutions.** Push for policies that require training, clear oversight processes, and fairness checks for tools in use.

- **Stay transparent with clients.** Explain that while AI may support care, ultimate responsibility rests with the clinician. This helps reinforce trust in the therapeutic relationship.

By making these practices part of daily work, clinicians preserve professional responsibility, strengthen transparency, and protect the ethical foundation of psychological care.

Defining the Line Between Support and Substitution

As AI tools grow more advanced, it is important to keep a clear line between support and substitution. AI can help with routine or analytic tasks such as summarizing records, flagging symptoms, or displaying data patterns. This can save time and allow clinicians to focus on direct care. But diagnosis, therapeutic formulation, interpretation, and the relational aspects of practice must remain firmly in human hands.

The risk is most visible in digital platforms where AI delivers interventions directly. Examples include CBT-based chatbots, symptom-tracking apps, or automated well-being "coaches." These tools may improve access to mental health support, particularly where resources are scarce, but they also risk replacing human connection. When this happens, the therapeutic alliance can weaken, and the line between professional care and automated advice becomes unclear. Clients may mistake algorithmic responses for clinical judgment, leading to confusion about who is truly responsible for care.

The same concern arises when institutions deploy AI as a cost-cutting measure, positioning it as a substitute for staff rather than a supplement. In these cases, ethical concerns intensify: Care risks can become more transactional, therapeutic nuance can be lost, and patients may be left without the empathy and contextual insight that define psychology.

Professional guidelines, including those from the American Psychological Association (APA) and other international organizations, emphasize that AI should enhance rather than replace human responsibilities. Transparency about what AI is doing, clear documentation of its influence, and reaffirming the clinician's authority are all essential for protecting the integrity of psychological practice.

What clinicians can do:

- **Define role boundaries clearly.** Let clients know which tasks AI supports (for example, summarizing progress notes) and which remain clinician-led (such as treatment planning).

- **Frame AI as assistive.** Make it clear, both in conversations and in documentation, that AI tools provide support but do not replace clinical expertise.

- **Guard against substitution creep.** Challenge situations where institutions or vendors suggest replacing core therapeutic functions with automation.

- **Use informed consent proactively.** Ensure clients understand how AI is used in their care and that human judgment remains central.

- **Advocate for guidelines.** Support professional standards that require transparency, human oversight, and client choice when AI is used.

By maintaining clear boundaries, clinicians can preserve therapeutic trust, safeguard professional responsibility, and ensure that AI strengthens rather than undermines the ethical foundation of psychological care.

★ **Key Takeaways:**

- **AI can assist but not replace.** While powerful at processing data, AI cannot replicate empathy, context, or moral judgment.

- **Over-reliance is a real danger.** Automation bias may cause clinicians to defer to outputs, risking misdiagnosis and loss of patient-centered care.

- **Oversight protects standards.** Clinicians and institutions must establish documentation, training, and review protocols to ensure AI remains a tool, not an authority.

- **Boundaries must be clear.** AI should support administrative and analytic tasks, while core responsibilities—diagnosis, interpretation, therapeutic alliance—must stay human-led.

- **Professional responsibility endures.** Ultimately, clinicians remain accountable for decisions. Upholding transparency and ethical practice safeguards both patients and the integrity of care.

Conclusion

AI has the potential to reshape psychological practice, but its integration must always be guided by ethics, oversight, and trust. This chapter has explored how regulatory frameworks lag behind technology, how privacy risks extend beyond traditional boundaries, how bias can reinforce systemic inequities, and how professional responsibilities remain central even with AI in the picture.

Across all of these issues, one principle is clear: AI should support, not replace, the human elements of care. Empathy, context, and ethical judgment cannot be handed off to algorithms. Developers, institutions, and clinicians all share responsibility for making sure AI is transparent, fair, and consistent with psychological values. Still, final accountability rests with human professionals.

For clinicians, the task is twofold: stay alert to the risks AI introduces and take an active role in shaping its use so that it strengthens, rather than weakens, therapeutic trust. This means asking critical questions about regulation and governance, protecting client privacy, recognizing bias, and setting clear boundaries for ethical practice.

Regulation and professional standards will continue to change, but patient care cannot wait. By engaging with AI critically and proactively, clinicians can help ensure these tools support the values of psychology, protecting dignity, equity, and trust in every setting.

Chapter 11: Frameworks for Evaluating AI Tools in Practice

The last chapter focused on the regulatory, ethical, legal, and bias-related challenges of using artificial intelligence (AI) in psychological practice. Setting standards is an important step, but it is not enough. We also need clear ways to test whether AI tools meet those standards. This is where evaluation frameworks become essential.

In this chapter, we look at several models that help psychologists, healthcare organizations, developers, and decision-makers judge whether an AI tool is safe, effective, fair, and ready to be used in real clinical settings. These frameworks turn broad principles into concrete criteria that can guide everyday choices. Whether you are selecting a tool, designing one, or reviewing its impact, these models provide a foundation for making informed and ethical decisions in a fast-changing field.

To make sense of the different approaches, we group them into four categories: frameworks for clinical and mental health assessment, models for evaluating organizational readiness, standards for ethical governance, and frameworks focused on education and human-AI interaction. Each perspective offers a way to decide not only whether an AI tool functions properly, but also whether it is ethically sound and appropriate for the context in which it is used.

Clinical and Mental Health Evaluation Frameworks

As the use of artificial intelligence in mental health care expands, new tools are being introduced to support diagnosis, therapy, and self-guided interventions. While these innovations create exciting possibilities, they also raise important questions about safety, effectiveness, ethical alignment, and readiness for real-world use. Traditional evaluation methods can fall short when applied to AI, so new frameworks have been created specifically for this purpose. In this section, we look at four such approaches: FAITA-Mental Health, AI for IMPACTS, the Thera-Turing Test, and the READI Framework. Each offers a different way to help ensure that AI systems are safe, effective, and consistent with professional standards.

FAITA-Mental Health

AI-driven tools in mental health—from chatbots to diagnostic aids—highlight the need for structured evaluation methods that focus on safety, ethics, and user-centered design. Unlike traditional therapeutic methods, these tools are often developed outside of clinical environments, with limited transparency about how they work, the data used to train them, or the logic behind their decisions. This creates a strong need for evaluation methods that are both clinically grounded and accessible to practitioners.

The FAITA-Mental Health framework was created to meet this need. It provides a way to evaluate AI tools in ways that safeguard autonomy, prevent harm, and align with clinical values. Mental health professionals, developers, policymakers, and institutions all share an interest in this process. Clinicians want to ensure patient safety and trust, developers seek practical guidance to meet professional standards, and institutions need reliable criteria for deciding whether tools are both ethical and effective.

Framework Overview

The Framework for AI Tool Assessment in Mental Health (FAITA-Mental Health) was created by Ashleigh Golden and Elias Aboujaoude as a systematic way to evaluate generative AI and other digital tools designed for mental health. It builds on earlier evaluation efforts, such as One Mind PsyberGuide, but expands and refines them to address the specific challenges posed by AI-powered interventions (Golden & Aboujaoude, 2024).

FAITA is organized into six domains, each with subdomains and a simple 0–2 scoring system:

- **Credibility** – Looks at whether the tool's goals are clearly defined, grounded in evidence, and whether it keeps users meaningfully engaged.

- **User Experience** – Reviews design features, personalization, the quality of interactivity, and whether feedback options are available.

- **User Agency** – Examines how well the tool protects privacy and autonomy, and whether it empowers users to make informed choices.

- **Equity and Inclusivity** – Considers cultural sensitivity and bias safeguards, ensuring the tool can serve diverse populations fairly.

- **Transparency** – Assesses how openly the tool discloses its ownership, business model, development process, and intended beneficiaries.

- **Safety and Crisis Management** – Evaluates whether the tool includes protocols for supporting users in crisis or at high risk, and whether those safeguards are reliable.

One strength of FAITA is that it can be used not only by clinicians, but also by developers and non-technical reviewers. By applying straightforward scores to each subdomain, it allows for clear, side-by-side comparisons of digital mental health tools, helping highlight both their strengths and areas for concern.

Application & Benefits

FAITA-Mental Health is especially useful for evaluating direct-to-consumer AI tools such as mental health chatbots or therapy assistants. It also provides guidance for developers who want to build tools that follow ethical and user-centered principles. Because it balances criteria like user agency, safety, and inclusivity with technical and clinical performance, it offers a well-rounded way to assess tools beyond traditional measures.

Golden and Aboujaoude demonstrated the framework's value by applying it to "OCD Coach," a generative AI assistant built with ChatGPT and marketed for obsessive-compulsive disorder. FAITA highlighted strengths such as the use of evidence-based interventions like Exposure and Response Prevention. At the same time, it revealed important gaps, including limited crisis response protocols and weak transparency around privacy and data governance. This example shows how

FAITA can serve both as a decision-making guide and as a tool for continuous improvement of mental health AI systems.

Overall, FAITA-Mental Health stands out for being clear, broad in scope, and accessible. It gives clinicians and organizations a structured way to cut through hype and marketing claims, replacing guesswork with evaluation grounded in ethics, usability, and clinical standards.

AI for IMPACTS Framework

In clinical practice, one of the biggest challenges is determining whether AI tools provide lasting value in real-world care. Accuracy and technical validation are important, but they are not enough on their own. Many evaluations focus narrowly on performance metrics while overlooking issues such as workflow integration, clinician trust, and long-term effectiveness. These are the factors that often determine whether a tool will succeed once it leaves the lab and enters daily use.

This gap highlights why tools that perform well in controlled studies may still fail to support care in real-world environments. For this reason, stakeholders such as clinicians, hospital administrators, procurement teams, and regulators need broader and more practical ways to evaluate AI. They need to know not only if a system works, but also whether it fits within existing processes, supports users effectively, and produces benefits that last over time.

The AI for IMPACTS framework was designed to meet this need. It offers a structured approach for assessing whether AI tools are not only technically sound but also practical, trustworthy, and sustainable in everyday clinical care.

Framework Overview

The AI for IMPACTS framework, developed by Jacob and colleagues (2025), offers a comprehensive way to evaluate AI-powered tools in clinical care. Built from a review and synthesis of 44 studies, it organizes evaluation into seven domains, each represented by a letter in the acronym IMPACTS:

- **Integration** – How smoothly the tool fits into existing workflows and connects with other systems.

- **Monitoring, Governance, and Accountability** – Oversight processes, ethical standards, informed consent, and clear lines of responsibility.

- **Performance and Quality Metrics** – Accuracy, reliability, fairness, and whether the tool's decisions can be explained.

- **Acceptability, Trust, and Training** – The degree of trust clinicians and patients have in the tool, and whether users receive proper training.

- **Cost and Economic Evaluation** – Financial considerations such as overall cost, return on investment, and efficient use of resources.

- **Technological Safety and Transparency** – Risk management, adherence to clinical standards, and openness about how the system makes decisions.

- **Scalability and Impact** – The ability of the tool to expand across settings or populations while maintaining long-term effectiveness.

Together, these domains cover 28 subcriteria that attempt to reflect the complexity of real-world healthcare. Unlike approaches that emphasize only technical performance, this framework highlights the importance of practical, ethical, and sustainable value in everyday use.

Application & Benefits

The AI for IMPACTS framework is especially useful for multidisciplinary teams that are deciding whether to purchase or adopt an AI tool in clinical practice. By focusing on issues such as integration, trust, oversight, and cost, it helps ensure that a system is not only technically effective but also clinically and ethically sound.

One of its key strengths is its sociotechnical design. The framework recognizes that technological performance and social context are deeply connected. It encourages organizations to ask practical questions early in the process, such as: Will this tool work in our specific setting? Who will be responsible for monitoring it? Can it be scaled in a way that is safe and responsible?

Because the framework is not tied to any single type of technology, it can be applied to a wide range of tools, including diagnostics, imaging, decision support, and patient monitoring. This flexibility makes it a valuable resource for comparative evaluations, internal audits, and procurement decisions. For healthcare organizations, it provides a structured way to deploy AI responsibly and sustainably.

Thera-Turing Test

As AI-based mental health chatbots, also called conversational agents, take on larger roles in providing mental health support, a critical gap has emerged: how do we know if these agents maintain the therapeutic quality, empathy, and fidelity to intervention methods that would be expected from human clinicians? This question is especially urgent in unsupervised settings, where users may rely on AI without a professional present.

Most existing evaluations focus on usability or outcomes like symptom reduction. Far fewer ask whether the actual conversations themselves meet the standards of safety, empathy, and therapeutic integrity expected in mental health care. Without this, there is no clear way to ensure that AI-delivered care aligns with professional ethics and safeguards.

This is a concern for multiple stakeholders. Clinicians need to know that these tools will not cause harm or lower therapeutic standards. Developers need structured feedback to improve model performance and reduce risk. Regulators need criteria for deciding whether tools are clinically ready and ethically sound. The Thera-Turing Test (TTT) was created to fill this gap.

Framework Overview

The Thera-Turing Test (TTT), proposed by Bunge and Desage (2025), is a structured way to evaluate AI-based mental health conversational agents. Unlike the original Turing Test, which focused on whether a computer could mimic human conversation in general, the TTT emphasizes therapeutic quality and competence. Its goal is to determine whether AI-generated conversations reflect the safety, clarity, and fidelity to practice expected from licensed clinicians.

At the heart of the TTT are independent clinician judges who review conversations without knowing whether they came from an AI agent or a human therapist. This blind review helps reduce bias and ensures greater objectivity.

The framework can be applied in several ways:

- **Transcript-only review** – Clinicians assess the quality of AI-generated therapy transcripts on their own.

- **Human vs. agent comparison** – AI session transcripts are directly compared with those of human therapists addressing the same topic.

- **Turn-by-turn comparison** – AI and therapists each respond to the same user input in alternating turns, allowing side-by-side evaluation.

- **Link to outcomes** – TTT ratings are compared with client outcomes to test whether higher-rated conversations actually lead to clinical benefit.

Conversations are scored across key therapeutic dimensions, including:

- Safety and appropriateness of content
- Ability to establish clarity and rapport
- Fidelity to evidence-based treatment methods
- Cultural sensitivity
- Quality of crisis response and referral practices

To strengthen reliability, the TTT also outlines protocols for selecting conversations (such as including both supportive and resistant users), training and blinding judges, and linking scores to readiness levels. These levels range from early-stage development to clinical trial testing to full deployment. The structure borrows from the American Psychological Association's Practicum Competency Assessment benchmarks, adapting them to conversational agents.

By focusing on therapeutic quality rather than technical fluency alone, the Thera-Turing Test provides a practical way to decide whether a chatbot or conversational agent is truly ready for clinical use.

Application and Benefits

The Thera-Turing Test (TTT) provides a clinically grounded way to evaluate AI-based conversational agents at different stages of development. It is especially valuable for assessing chatbots designed to operate with little or no human supervision. By focusing on therapeutic process and quality—rather than just technical fluency—it helps determine whether an AI agent can realistically support users' mental health needs.

For developers, the TTT offers structured, psychometrically informed feedback that can guide design and refinement. For clinicians and organizations, TTT scores provide a practical basis for deciding whether a conversational agent is ready for use in real-world settings. Because the framework includes multiple assessment strategies, it can be applied early in development to shape design, or later on to support clinical trials and marketing claims.

Most importantly, the TTT sets a gold standard for evaluating mental health chatbots. It gives the field a scalable way to measure safety, therapeutic integrity, and user alignment. In doing so, it helps ensure that digital therapeutics are held to the same expectations as human clinicians, strengthening both accountability and trust in AI-supported care.

READI Framework

Despite rapid adoption, many AI tools are introduced into mental health care without adequate review of their readiness, safety, or fit for real-world practice. While there is growing attention to oversight and ethics, most existing frameworks focus narrowly on one area—such as algorithmic fairness or safety—without addressing the broader clinical and social context. As a result, tools may reach patients with little scrutiny of whether they can be safely used, responsibly integrated, or adapted to complex care environments. Risks include mishandling of sensitive disclosures, biased conversational patterns, or a lack of safeguards against misuse and overuse.

Clinicians, administrators, developers, and healthcare organizations all need a practical, standardized way to evaluate whether an AI tool is safe, effective, ethical, and truly ready for deployment. The READI Framework—short for *Readiness Evaluation for AI-Mental Health Deployment and Implementation*—was developed to meet this need. It provides a structured approach to assessing AI tools not just for technical quality, but for their clinical, ethical, and contextual appropriateness in everyday practice.

Framework Overview

Developed at Stanford University's Institute for Human-Centered Artificial Intelligence, the READI Framework provides a six-domain model for assessing whether AI tools are truly ready for use in mental health care. It adapts lessons from health, ethics, AI governance, and implementation science, while focusing specifically on the clinical and psychosocial challenges unique to mental health technologies (Stade et al., 2025).

The six domains are:

- **Safety** – Ensures the tool prevents harm, responds appropriately in crises, and avoids encouraging unhealthy or harmful behaviors.

- **Privacy and Confidentiality** – Looks at data protections, transparency in how information is used, and the level of control users have over their personal data.

- **Equity** – Examines how well the tool serves diverse populations, whether it has been tested for bias, and how cultural and demographic factors are addressed.

- **Effectiveness** – Requires clear evidence that the tool improves clinical outcomes such as symptom reduction or overall well-being, based on rigorous evaluation.

- **Engagement** – Considers whether the tool encourages healthy, ongoing use without creating dependency or misuse.

- **Implementation** – Reviews how practical the tool is to deploy in real-world settings, including cost, compatibility with existing systems, regulatory compliance, and ease of use for both clinicians and patients.

Rather than producing a simple score, the READI Framework guides structured, context-specific judgments about whether an AI tool is clinically appropriate. It offers practical criteria that can inform both the development of new tools and decisions about adoption in care settings.

Application and Benefits

The READI Framework works well at different stages of AI tool development and use—whether a tool is in its early pilot phase, being refined for wider testing, or already close to clinical rollout. It gives institutional review boards, procurement teams, developers, and clinical supervisors a structured way to check readiness and spot warning signs. Unlike many general healthcare or consumer app frameworks, READI includes unique elements such as engagement and implementation, recognizing that mental health tools bring distinct challenges and risks.

One of READI's major strengths is that it avoids a "one-size-fits-all" formula. Instead, it emphasizes transparency, fit within real-world contexts, and awareness of potential risks. This flexibility makes it valuable for different decision-makers—whether clinicians, administrators, or developers—who want to build trust and ensure responsible adoption. As AI applications for mental health continue to expand, READI offers a dynamic, ethics-driven approach rooted in human-centered design and implementation science.

Taken together, the four frameworks discussed—FAITA, AI for IMPACTS, the Thera-Turing Test, and READI—provide complementary ways of looking at clinical AI. They range from evaluating therapeutic quality and user experience to assessing system readiness and institutional adoption. Each framework highlights a different piece of the puzzle, helping professionals make informed, ethical choices in a rapidly evolving landscape.

While clinical frameworks are essential for checking whether tools are safe, ethical, and effective, successful use in practice often depends on more than clinical performance alone. Institutions also need to examine their own ability to adopt, oversee, and sustain AI responsibly. The next section turns to organizational readiness frameworks, which help healthcare systems evaluate their strategy, infrastructure, and culture for AI integration.

Organizational Readiness and Capability Models

As organizations move from simply exploring artificial intelligence to putting it into practice, their success depends on more than just the technology itself. Readiness must also exist at the strategic level (clear goals and leadership support), the infrastructural level (the systems and resources needed to make it work), and the cultural level (whether staff and stakeholders are prepared to adopt it responsibly).

This section introduces three frameworks that help organizations evaluate their readiness from different perspectives:

- **The AI Readiness Index** – A practical tool, especially helpful for small and mid-sized organizations, to evaluate how prepared they are to adopt AI.

- **The AI Capability Assessment Model** – A maturity model that looks at how organizations can build and scale their capabilities, with a focus on governance and structured growth.

- **The AI Readiness Framework** – An approach that emphasizes the human and social side of digital transformation, not just the technical components.

Together, these models provide organizations with different but complementary ways to determine whether they are truly ready to bring AI into practice responsibly.

AI Readiness Index (AIRI)

Many organizations are eager to use artificial intelligence (AI) to improve efficiency and innovation, but not all are fully prepared to implement these tools. This is especially true for small and medium-sized organizations, where interest in AI often grows faster than the infrastructure, staffing, or governance needed to support it. Challenges may include limited technical systems, lack of staff training, or uncertainty about policies and oversight.

The AI Readiness Index (AIRI) was created to address this gap. Originally developed by AI Singapore and supported in academic research, AIRI provides organizations with a structured way to assess whether they have the foundational capabilities required to use AI responsibly. It goes beyond traditional digital readiness checklists by also examining factors such as employee attitudes toward AI, governance maturity, and organizational culture.

Importantly, AIRI does not just highlight gaps—it also provides a roadmap for planning adoption strategies that are realistic for an organization's size, resources, and sector. In doing so, it helps organizations move from enthusiasm to practical, responsible implementation.

Framework Overview

The AI Readiness Index (AIRI) evaluates how prepared an organization is to adopt AI by looking at five major areas, broken down into twelve specific dimensions. Each area is rated on a scale from 0 to 3, which then places the organization into one of four overall readiness levels: AI Unaware, AI Aware, AI Ready, or AI Competent.

The five pillars include:

1. Organizational Readiness

- *AI Literacy* – Do staff and leaders understand the basics of AI?
- *AI Talent* – Does the organization have access to people with the right skills?
- *Management Support* – Are leaders actively supporting AI adoption?
- *Employee Acceptance* – Are staff open to using AI in their work?
- *Experimentation Culture* – Is there room to test new ideas safely?

2. Ethical and Governance Readiness

- *AI Governance* – Are there clear rules and oversight for AI use?

- *AI Risk Control* – Are risks identified and managed responsibly?

3. Business Value Readiness

- *Use Case Identification* – Has the organization identified meaningful, practical uses for AI?

4. Data Readiness

- *Data Quality* – Is the data accurate and reliable?
- *Reference Data* – Are there consistent standards for how data is used?

5. Infrastructure Readiness

- *Machine Learning Infrastructure* – Are the technical systems in place to run AI tools?
- *Data Infrastructure* – Is the organization's data stored and organized in ways that support AI use?

The AIRI assessment generates a heatmap that shows strengths and vulnerabilities across these domains. Through its online platform, organizations receive a tailored report with an overall readiness score, comparisons with similar organizations, and specific recommendations for improvement.

Application and Benefits

In a 2024 study, Budiraharjo and colleagues applied the AIRI framework to six small and medium-sized businesses (SMEs) in Indonesia across a range of industries. The results showed how AIRI can reveal not just an organization's current readiness level, but also the specific steps needed for improvement.

For example:

- Some organizations, like CV Putra Wijaya Konstruksi, were classified as *AI Aware,* meaning they had a basic foundation and clear goals but still needed to strengthen infrastructure and skills before moving forward.

- Others were classified as *AI Unaware,* indicating they needed to start with fundamentals such as staff education, governance, and better data systems before AI adoption could realistically happen.

Each organization received both general guidance and targeted recommendations, which included training in AI ethics, improving data management practices, and setting aside budget for technology infrastructure.

Importantly, participants in the study agreed that the AIRI framework was comprehensive, accurate, and easy to act on. This highlights one of AIRI's greatest strengths: it captures both the technical side (such as data systems and infrastructure) and the human side (such as leadership support and organizational culture).

Although this study focused on SMEs, AIRI's approach is scalable. It can help larger institutions, including healthcare and mental health organizations, evaluate their preparedness for AI in a balanced way—considering not only the technology itself but also the culture, ethics, and governance needed for safe and responsible implementation.

AI-CAM (AI Capability Assessment Model)

As organizations expand their use of AI, many realize they don't yet have a clear picture of what successful adoption really requires. This is not only about technology; it's about whether the organization has the right skills, ethical safeguards, and alignment with overall goals.

Some organizations fall into "AI-washing," claiming they use AI but lacking the necessary components, such as reliable data, skilled staff, or proper oversight. Leaders may be uncertain about how to:

- Connect AI projects to their organization's larger mission,
- Decide which teams need which new skills, or
- Evaluate the ethical and regulatory risks of these tools.

The AI-CAM (AI Capability Assessment Model) was created to address these challenges. It provides a structured maturity model that organizations can use to measure how ready they are to adopt AI responsibly. AI-CAM helps leaders and teams:

- Identify their current level of capability,
- Understand what growth and progress look like, and
- Close the gaps that may prevent them from using AI effectively and ethically.

Framework Overview

The AI Capability Assessment Model (AI-CAM), created by Butler, Espinoza-Limón, and Seppälä, is a maturity model that helps organizations understand how ready they are to use AI responsibly. It evaluates progress across seven key areas:

- **Business** – How well AI adoption is tied to the organization's overall goals and vision.
- **Data** – The quality, structure, and governance of data across the organization.
- **Technology** – The platforms and tools available to develop, deploy, and maintain AI.
- **Organization** – How prepared the internal culture and structures are to integrate AI.
- **AI Skills** – The workforce's knowledge of technologies that support AI, including both technical and applied skills.
- **Ethics and Risk** – The policies and practices in place to address ethical issues, regulatory requirements, and reputational risks.
- **Use Cases (Applications)** – The ability to identify, prioritize, and evaluate AI initiatives that are not only feasible but also ethical and aligned with strategic goals.

AI-CAM uses five levels of maturity to show where an organization stands:

- **Initial** – Early experimentation, with little connection to strategy and limited skills.
- **R&D** – Pilot projects have started, and data and governance needs are becoming clearer.
- **Strategic** – Pilots are expanding, with clearer strategy, better data practices, and growing workforce skills.
- **Defined** – AI is being deployed in production, with strong integration of skills, ethics, and measurable outcomes.
- **Quantitatively Managed** – AI is embedded across the organization, with systems for continuous improvement and active risk governance.

In addition, the AI Capabilities Matrix (AI-CM) outlines the knowledge and skills expected at different roles and proficiency levels. It covers executives, project leads, and technical staff, defining what's needed at basic, advanced, and expert levels. The matrix is grounded in current research and job market demands, spanning areas such as machine learning, system administration, knowledge representation, and ethics.

Together, AI-CAM and AI-CM help organizations take a structured, step-by-step approach to AI adoption, helping to ensure they not only pursue innovation but also do so responsibly, with the right infrastructure, skills, and safeguards in place.

Application and Benefits

The AI Capability Assessment Model (AI-CAM) is especially useful for large organizations that want to move past small pilot projects and scale AI responsibly. It gives leaders a clear way to evaluate their current readiness, spot gaps in skills or infrastructure, and plan the investments needed to grow—whether in governance, data, or workforce development.

A key strength of AI-CAM is that it provides specific, observable criteria for each stage of maturity. This makes it possible for organizations to benchmark progress and track improvements over time.

Because the model is not tied to any one sector or type of AI, it can be applied broadly. Organizations can use it to:

- Assess their own internal capabilities,
- Evaluate the readiness of external vendors, or
- Design targeted training programs for staff using the AI Capabilities Matrix (AI-CM).

Importantly, AI-CAM places ethical oversight at the center of its framework. Rather than treating ethics as a compliance requirement tacked on at the end, it highlights ongoing monitoring of bias, privacy, and social impact as essential to scaling AI. This ensures that as organizations expand their use of AI, they do so responsibly and sustainably, without compromising trust or equity.

AI Readiness Framework

Adopting AI is never just about installing new technology. It happens within the larger systems of an organization—shaped by its people, daily workflows, and cultural attitudes. Many organizations underestimate how much alignment is needed across these areas. Without it, AI projects often end up underused, poorly integrated, or misaligned with broader goals.

The AI Readiness Framework was created to evaluate this full picture. Instead of looking only at technical capability, it examines how well the social and organizational context supports AI adoption.

This framework is especially useful for executives, digital transformation leaders, and IT strategists who need to know whether their organization can truly support, scale, and sustain AI. It helps answer not just "Can we implement AI?" but the deeper question: "Is our organization prepared to use AI effectively and responsibly?"

Framework Overview

The AI Readiness Framework, developed by Jonny Holmström, is designed to help organizations see how prepared they really are to use AI in ways that support genuine digital transformation. Its core idea is that success with AI depends not only on technology but also on people, structures, and culture.

The framework looks at readiness across four connected areas:

- **Technologies** – How strong the organization's AI infrastructure, data systems, and tools are now and in the future.
- **Activities** – How well AI supports everyday workflows and value-generating tasks, and whether the organization is open to reshaping those processes.
- **Boundaries** – How AI changes team dynamics, workflows, or even how the organization engages with clients and partners.
- **Goals** – How clearly AI adoption aligns with leadership vision, organizational strategy, and cultural readiness for change.

Each area is scored from 0 to 4 in two ways:

- **Current capability** (what exists today).
- **Future potential** (what the organization is planning or aspiring to).

For example, a clinic might have a strong technology system in place now but lack a long-term vision for how AI will support patient care. These kinds of mismatches highlight where leadership should focus to ensure AI adoption is not only possible but also sustainable.

Application and Benefits

The AI Readiness Framework is a hands-on tool organizations can use in workshops or self-assessments to understand their capacity for AI adoption. It helps leadership teams map out current strengths, identify weak spots, and plan where to focus next.

With this framework, organizations can:

- **Benchmark their AI maturity** across different areas, from technology to culture.

- **Spot priority gaps** where investment in skills, governance, or infrastructure is needed.

- **Uncover misalignments** between ambitious AI goals and the organization's actual readiness to support them.

- **Support cultural change**, encouraging teams to adapt workflows and mindsets as AI becomes part of daily practice.

For example, in a pilot with an insurance company exploring chatbots and analytics, leaders discovered they had strong technical capacity in data analysis but were hesitant to use AI in customer-facing roles. By making these misalignments visible, the framework helped them decide where training, communication, and change management should come first.

What makes this model unique is its sociotechnical perspective—it doesn't look at AI as just another IT project. Instead, it emphasizes that goals, structures, and workflows all need to evolve together for AI to bring real value. This makes it a practical planning and communication tool for executives working across departments and stakeholder groups.

Still, being "ready" is only the first step. Once AI is embedded in critical systems, organizations must also set up strong governance structures to manage ethical risks, meet regulatory requirements, and ensure accountability. The next section introduces governance-focused frameworks that turn high-level principles into operational safeguards and standards organizations can follow.

Ethical Risk & Governance Frameworks

As artificial intelligence becomes more deeply woven into healthcare, public services, and private organizations, the need for strong ethical oversight and governance is greater than ever. The speed of AI adoption has often outpaced the creation of clear rules and standards, leaving professionals and institutions under pressure not only to innovate, but also to prove that their systems are safe, fair, accountable, and aligned with human values.

This section introduces three governance-oriented frameworks that can help:

- **ISO/IEC AI Standards (JTC 1/SC 42):** A set of international guidelines that provide structure for managing risks across the AI lifecycle, ensuring systems work together, and promoting ethical alignment.

- **Responsible AI Question Bank:** A practical tool that translates broad ethical principles into specific questions organizations can ask at different stages of AI development and use. It helps ensure accountability is considered across roles and responsibilities.

- **AI Governance and QA Framework (Pharmaceutical Model):** Originally designed for the pharmaceutical industry, this framework shows how AI can be integrated into existing regulatory and quality assurance systems—ensuring compliance while maintaining patient safety.

Together, these approaches give organizations tools to embed governance into AI from the start. They not only reduce risks but also help build systems that are transparent, auditable, and trustworthy by design.

ISO/IEC JTC 1/SC 42 Technical Subcommittee

As artificial intelligence continues to advance, organizations are under growing pressure to make sure the systems they adopt are not only innovative but also safe, transparent, and accountable. A major challenge is that many industries and countries lack a consistent set of rules or standards for managing ethical risks, protecting data, and ensuring systems work together. This leaves companies, regulators, and technology providers struggling to navigate fragmented expectations and uncertainty.

To address this, two major international bodies—the International Organization for Standardization (ISO) and the International Electrotechnical Commission (IEC)— created a special subcommittee called ISO/IEC JTC 1/SC 42. This group develops international standards that set out common language, shared definitions, and best practices for AI. These standards cover areas such as risk management, data governance, and ethical safeguards, helping organizations around the world move toward trustworthy and interoperable AI practices.

Mandate and Focus Areas

ISO/IEC JTC 1/SC 42 was launched in 2018 as the first international standards body focused entirely on artificial intelligence. Its mission is to create international standards and technical reports that promote safe, responsible, and consistent development of AI technologies worldwide (ISO/IEC, 2023).

SC 42 organizes its work into several key areas:

- **Foundational Standards**: Establish definitions, taxonomies, and concepts for AI and machine learning (e.g., ISO/IEC 22989:2022).

- **Data Standards**: Provide guidance on data quality, provenance, bias reduction, and privacy (e.g., ISO/IEC 5259 series).

- **Risk and Trustworthiness**: Address robustness, explainability, safety, and security (e.g., ISO/IEC 23894:2023, a risk management standard for AI).

- **Lifecycle Governance**: Define requirements and best practices for managing AI systems throughout their lifecycle, including impact assessments, monitoring, and retirement.

- **Ethical and Societal Considerations**: Cover accountability, transparency, fairness, and human oversight.

SC 42 also works with other ISO/IEC committees, including those focused on cybersecurity, cloud computing, and health informatics. It collaborates with national standards bodies and international regulators. While the standards it produces are voluntary, they are increasingly shaping national policies and emerging regulations, such as the EU AI Act and Canada's AIDA.

Application and Benefits

SC 42's standards give organizations a shared global reference for developing and managing AI systems. The standards are modular, which means they can be applied based on context, such as creating clinical decision support tools in healthcare or deploying generative AI in consumer applications.

- **For technology developers**, SC 42 standards guide internal governance, product design, and preparation for international regulatory requirements.

- **For regulators and auditors**, they provide a structured way to assess AI risk and compliance.

- **For enterprises and institutions**, they help build ethical and operational safeguards directly into technical workflows, moving responsible AI from aspiration to real-world practice.

By focusing on lifecycle governance, collaboration across disciplines, and trustworthiness, SC 42 helps shape a more consistent, transparent, and ethically grounded global AI ecosystem.

Responsible AI Question Bank

As AI becomes part of everyday products, services, and decision-making systems, organizations are under pressure to show that their tools are ethical, trustworthy, and aligned with new regulations and social expectations. Traditional risk assessments are often not enough. They may lack structure or fail to address the broad ethical principles that AI governance now requires. Many teams also struggle to take values like fairness, transparency, or accountability and apply them in day-to-day work across roles, departments, and development stages.

The Responsible AI Question Bank was created to close this gap. It turns abstract ethical principles into clear, practical questions that can be built into AI risk assessments, development workflows, and governance processes. The tool is designed to help organizations bring their practices into alignment with international standards, including the EU AI Act, ISO/IEC 42001, and the OECD AI Principles.

Framework Overview

The **Responsible AI Question Bank**, developed by Lee, Hovy, Luccioni, and Mittelstadt (2024), is a structured set of 245 questions designed to guide thorough, ethics-informed reviews of AI systems. It is built around eight core AI ethics principles drawn from academic, government, and international sources:

- Human agency and oversight
- Technical robustness and safety
- Privacy and data governance
- Transparency
- Diversity, non-discrimination, and fairness
- Societal and environmental well-being
- Accountability
- Contestability and redress

Each principle is expressed through questions that vary in depth and intended audience. The framework organizes these into three levels:

- **Level 1** – General awareness, aimed at executives and organizational leadership
- **Level 2** – Organizational practices, for project managers and risk teams
- **Level 3** – Technical detail, tailored for developers, designers, and engineers

The question bank is also aligned with the AI lifecycle, addressing planning, development, deployment, and post-deployment monitoring. This helps organizations apply ethical thinking across the entire process, rather than limiting it to a single compliance step.

The tool can be used on its own or paired with other frameworks. It has been mapped to the EU AI Act and ISO/IEC 42001, making it useful for legal and standards compliance. By asking questions instead of providing yes-or-no checklists, it encourages reflection, dialogue, and ongoing accountability rather than surface-level compliance (Lee et al., 2024).

Application and Benefits

The **Responsible AI Question Bank** gives organizations a practical but thorough way to run internal reviews, design governance processes, and document compliance. Because it is modular and role-specific, it can be applied in many situations, such as:

- Supporting internal governance boards or AI ethics committees
- Helping with procurement reviews of third-party AI tools
- Guiding responsible innovation during system design
- Aligning development with upcoming AI regulations

In case studies shared by the authors, the tool was piloted in global technology companies and public institutions. Participants said the structured, multi-level questions helped them uncover blind spots, reduce uncertainty, and create shared understanding between legal, technical, and managerial staff.

The question bank does not generate a score or a pass-fail outcome. Instead, it serves as a framework to build a culture of ethical reflection and responsibility throughout the AI development process. It is particularly helpful for organizations preparing for formal audits or external assurance reviews.

AI Governance and QA Framework

In highly regulated fields such as pharmaceuticals, using artificial intelligence comes with challenges that extend beyond the technical side. AI must be built into existing Good Automated Manufacturing Practice (GxP) systems while still meeting strict standards for quality, traceability, and compliance. Traditional quality assurance approaches often struggle with AI because these tools evolve through iterative development and show probabilistic behavior, rather than fixed outcomes.

The AI Governance and QA Framework was created to meet these needs. It offers a structured way to design, validate, and maintain AI solutions within GxP environments. The framework highlights the importance of a comprehensive governance model that reflects the unique features of AI while also maintaining the high-quality standards required in the pharmaceutical industry.

Framework Overview

Developed by Elias Altrabsheh, Martin Heitmann, and Albert Lochbronner, this framework provides a comprehensive model for AI governance and quality assurance in pharmaceutical environments. It is built around an AI Quality Assurance Master Plan, which covers six main areas:

- **Corporate Culture**: Building a mindset that supports change, learning, and responsible AI adoption.

- **People and Skills**: Training and aligning staff across departments so they have the right competencies for AI development and quality assurance.

- **AI Governance Processes**: Using iterative processes that reflect the evolving nature of AI while maintaining compliance with GxP standards.

- **Information, Data, and Sources**: Ensuring data quality and fitness for purpose, since data is the foundation of all AI systems.

- **Software and Algorithms**: Evaluating both in-house and third-party algorithms for complexity and suitability in production environments.

- **Services, Infrastructure, and Platforms**: Establishing reliable infrastructure that can manage large data volumes and support real-time AI operations.

The framework also sets out a three-phase lifecycle for managing AI solutions:

- **Project Initiation and Initial GxP Assessment**: Reviewing feasibility and compliance of proposed AI applications.

- **Development, Quality Assurance, and Productive Operation**: Following an iterative process that keeps development and quality assurance separate to maintain compliance and objectivity.

- **Product Discontinuation and Retirement**: Planning the safe retirement of AI systems while preserving data integrity and meeting regulatory requirements.

This model helps organizations integrate AI in ways that are both flexible and compliant with pharmaceutical quality standards (Altrabsheh, Heitmann, & Lochbronner, 2022).

Application and Benefits

The AI Governance and QA Framework provides a practical way for pharmaceutical organizations to adopt AI responsibly. It connects AI development with established quality management systems and regulatory requirements, making sure that innovation does not come at the expense of compliance. The framework supports four main goals:

- **Regulatory Compliance**: Ensuring AI applications meet GxP standards and are ready for regulatory review.

- **Risk Mitigation**: Identifying and addressing risks tied to the use of AI in sensitive processes.

- **Operational Efficiency**: Improving the development and validation of AI tools through structured governance and quality practices.

- **Sustainable Innovation**: Encouraging ongoing improvement and the ability to scale AI solutions across the pharmaceutical lifecycle.

By offering this structure, the framework helps organizations unlock the benefits of AI while maintaining the highest standards of safety, quality, and compliance.

Expanding the Governance Lens

While the pharmaceutical field requires strict technical and regulatory oversight, other areas such as education highlight different priorities. Here, the focus is less on compliance and more on how people experience and trust AI. When AI tools interact directly with learners, the most important issues become transparency, user comprehension, engagement, and maintaining personal autonomy.

This next section looks at frameworks that evaluate the relationship between people and AI, with an emphasis on building systems that are clear, trustworthy, and responsive to user needs.

Education and Human-AI Interaction Frameworks

As artificial intelligence tools become more common in education, new challenges arise in how students, teachers, and institutions interact with them. These challenges are not just technical. They also involve questions about trust, fairness, communication, and whether the tools are being used in appropriate ways. Traditional evaluation methods often miss this human side of AI, overlooking how people learn with, adapt to, or are influenced by these systems.

To help fill these gaps, researchers have developed new frameworks that focus on the human and educational aspects of AI. These models look beyond what AI systems can do and instead ask how people understand, guide, and evaluate their behavior. This section introduces three such frameworks: the Artificial Intelligence Assessment Scale (AIAS), the SUDO framework, and the Human-Aware AI (HAAI) framework. Each provides a different perspective on how to design, assess, or govern AI in ways that respect human expectations and support meaningful learning.

Taken together, these approaches encourage us to see AI in education not only as a set of tools to manage, but as systems that must be understood and integrated with human judgment.

Artificial Intelligence Assessment Scale (AIAS)

As generative AI becomes part of classrooms and educational settings, educators and institutions face a difficult question: how can these tools be used in ways that are educationally sound, ethically responsible, and consistent with academic integrity? Many schools do not yet have clear policies, teachers apply different rules, and students often receive mixed or confusing guidance about when and how AI can be used. This lack of clarity can lead to anxiety, unfairness, and confusion, especially in exams and assignments where tools like ChatGPT are easy to access.

The Artificial Intelligence Assessment Scale (AIAS) was developed to address these issues. Instead of viewing AI as automatically harmful or banning it completely, AIAS provides a structured and scalable way to integrate AI into learning assessments. Its goal is to encourage ethical use while also fostering critical thinking and student responsibility.

Framework Overview

Developed by Perkins, Furze, Roe, and MacVaugh (2024), the Artificial Intelligence Assessment Scale (AIAS) is a five-point system that helps set clear boundaries for how generative AI can be used in academic work. The scale ranges from a complete ban on AI to full collaboration with it. This gives both educators and students clarity about what is acceptable and ties those boundaries to learning goals and academic integrity.

The five levels are:

- **Level 1: No AI Use.** All work must be completed without AI support.

- **Level 2: AI-Assisted Idea Generation and Structuring.** Students may use AI for brainstorming or outlining, but the final submission cannot include AI-generated text.

- **Level 3: AI-Assisted Editing.** AI may be used for grammar, language, or formatting, but students must turn in both the original and the edited versions.

- **Level 4: AI Task Completion with Human Evaluation.** AI can complete specific tasks, but students are responsible for reviewing and critically assessing its outputs.

- **Level 5: Full AI Collaboration.** AI may be used openly at every stage, but students remain accountable for the final product.

The scale is designed to be flexible. Educators can choose levels based on their teaching goals, and institutions can adapt it to match policies or disciplinary needs (Perkins, Furze, Roe, & MacVaugh, 2024).

Application and Benefits

The AIAS is useful for both institutional policy and classroom practice. It gives educators a clear way to set expectations, redesign assignments to include AI thoughtfully, and help students use generative tools responsibly. The framework helps address several key challenges:

- **Academic Integrity.** By clearly outlining what is and is not allowed, the scale reduces confusion and helps prevent misconduct.

- **Skill Development.** It builds digital literacy, ethical reasoning, and critical thinking by linking AI use to learning goals.

- **Equity and Transparency.** Clear rules ensure students from different backgrounds understand expectations, which supports fairness in how AI is used.

- **Institutional Alignment.** Schools and universities can use the scale as a foundation for policies, course design, and faculty training.

By moving away from the idea of simply banning AI, the AIAS encourages constructive integration. It helps students learn *about* AI and learn *with* AI, preparing them for a future where working alongside these tools will be the norm.

SUDO Framework

Artificial intelligence systems are often used on data that look very different from the data they were trained on. These real-world, or "data in the wild," situations can cause problems. One of the biggest issues is that ground-truth labels—verified outcomes that tell us whether AI predictions are correct—are often missing. In education, this is especially common. Labels for things like student performance, engagement, or risk are not always available, making it hard to check how well AI tools really work. Without these labels, schools may not be able to properly audit models, identify bias, or know if predictions are reliable.

The SUDO framework (pseudo-label discrepancy) was developed to help solve this problem. It provides a way to evaluate AI systems even when ground-truth labels are not available. By doing so, it helps educators and other stakeholders:

- Detect unreliable predictions
- Choose models that are more trustworthy
- Assess potential algorithmic bias

Although it was first created for clinical data, the SUDO approach can also be applied to education and other areas where verified outcomes are hard to get.

Framework Overview

The SUDO framework uses a five-step process to test how reliable an AI system is when used in real-world conditions where ground-truth labels may not be available (Kiyasseh, Cohen, Jiang, & Altieri, 2024):

1. **Run the AI on real-world, unlabeled data.** The system produces probability scores for each case.

2. **Group the results into ranges.** These ranges, or "quantile-based intervals" (such as deciles), organize the probability scores into more manageable categories.

3. **Create pseudo-labels.** Within each interval, assign provisional class labels to sampled data points and compare them with known examples from the opposite class.

4. **Train and test classifiers.** Use these pseudo-labeled datasets to train classifiers and then repeat the process with different pseudo-labels for comparison.

5. **Measure discrepancies.** Compare performance on held-out ground-truth data to calculate the "pseudo-label discrepancy." This number shows how reliable the AI's predictions are for each interval.

A lower discrepancy means the model is likely mixing classes incorrectly and is less reliable. A higher discrepancy suggests cleaner separation between classes and greater predictive trustworthiness.

One of SUDO's strengths is that it does not rely only on the AI's self-reported confidence scores, which are often misleading. This makes it particularly valuable for identifying when an AI system is overly confident in its predictions.

Application and Benefits

The SUDO framework offers a scalable way to evaluate AI systems when labeled data are not available or are too costly to obtain. It is especially useful in high-stakes areas such as education, where prediction accuracy must be trusted but constant human annotation is not realistic.

Key benefits include:

- Spotting unreliable predictions so they can be reviewed before they affect students or important decisions.
- Checking for algorithmic bias across different demographic or institutional groups, even without labeled data.
- Choosing AI models based on how stable their pseudo-labeling is, rather than assuming performance will carry over to every new setting.

SUDO is flexible across data types. It can be applied to text, images, and structured records, making it useful for learning analytics, predictive student support, or AI-based tutoring tools. It also helps institutions monitor AI systems after deployment, offering a way to trigger audits, retraining, or adjustments when data patterns change over time.

Although SUDO was first tested in clinical fields, its core principles fit well with the needs of educational AI governance. By building SUDO into institutional review processes, schools and universities can strengthen fairness, reliability, and transparency, even when they have limited labeled data to work with.

Human-Aware AI (HAAI) Framework

As artificial intelligence tools become more common and more powerful, one of the biggest challenges is making sure they interact with people in ways that feel clear, trustworthy, and consistent with human expectations. In sensitive areas such as healthcare, education, or decision support, problems arise when an AI system's behavior does not match what the user expects. This can lead to frustration, loss of trust, and weaker outcomes.

Issues like explainability or fairness are often treated separately, but they stem from the same root problem: a mismatch between what people think the system will do and what it actually does.

The Human-Aware AI (HAAI) framework was created to address this challenge. It offers a structured way to model human–AI interaction as an ongoing process shaped by differences in beliefs, abilities, and perspectives. Instead of focusing only on isolated technical fixes, HAAI provides a unified approach for understanding and improving how people and AI work together.

Framework Overview

Developed by Sarath Sreedharan (2023), the Human-Aware AI (HAAI) framework draws on insights from psychology, human–computer interaction, and decision-making research. It starts with a simple but powerful idea: every interaction between people and AI involves some level of mismatched expectations. Users come with assumptions about what an AI can do, how it reasons, or how it communicates. These assumptions are often incomplete or inaccurate, which can cause confusion when the AI behaves differently than expected.

HAAI describes three main types of mismatches, called asymmetries:

- **Knowledge Asymmetry**: when people and AI hold different beliefs about the world or about the AI's capabilities.

- **Inferential Asymmetry**: when they use different reasoning processes to decide on plans or goals.

- **Vocabulary Asymmetry**: when they rely on different language or concepts to describe a task.

These gaps lead to two major challenges in human–AI interaction:

- **Explanation Generation**: helping people understand why the AI made a certain choice, often by clarifying hidden assumptions.

- **Value Alignment**: helping the AI infer what the person really intended, even when the request was vague or incomplete.

HAAI views explanation as a corrective process. If a user is surprised by the AI's action, the AI should pinpoint the misunderstanding and offer clarification. For example, if a robot heats water in a microwave instead of using a kettle, it should explain that it cannot use a kettle safely, something the user may not have known.

For value alignment, the process works in reverse. The AI must reason about what the user meant. If the user asked the robot to make tea but did not specify a brand, the AI should be able to guess that preference, or ask for clarification, so the outcome matches the user's intent (Sreedharan, 2023).

Application and Benefits

The Human-Aware AI (HAAI) framework provides a structured way to design and evaluate AI systems that interact directly with people. It is particularly helpful in fields where trust and clear communication are essential.

Key benefits include:

- **Unified Theory of Interaction**: Explains why problems of trust or usability often arise and does so across different domains.

- **Actionable Design Insights**: Guides developers to identify and address the main mismatches between users and AI systems.

- **Support for Ongoing Clarification**: Encourages AI systems to notice when expectations diverge and adjust their responses in real time.

- **Foundation for Human-Centered Metrics**: Expands evaluation beyond technical accuracy to include alignment with user expectations and overall satisfaction.

The HAAI framework has already influenced work in areas such as explanation methods, preference identification, and trust-building. Its central contribution is a pathway toward AI systems that are not only effective but also sensitive to how people think, communicate, and make decisions.

Conclusion

The integration of AI into mental health and psychological practice cannot be guided by intuition or marketing claims alone. It requires structured evaluation across multiple perspectives. The frameworks presented in this chapter—FAITA, AI for IMPACTS, the Thera-Turing Test, READI, AIRI, AI-CAM, and others—highlight that success depends on more than technical accuracy. True readiness involves ethical grounding, clinical alignment, and fit within real-world contexts.

No single framework captures every dimension, but taken together, they offer psychologists and organizations the tools to ask the right questions and make informed choices. These models support the critical task of determining whether an AI tool is not only functional but also safe, trustworthy, and appropriate for client care.

As AI continues to evolve, these approaches will help ensure that adoption is guided by both innovation and integrity.

Chapters 10 and 11 set the stage by outlining the ethical boundaries and practical frameworks needed to guide the use of AI in psychological practice. We looked at how core values such as privacy, fairness, and clinical judgment intersect with AI, and how structured models can help assess readiness, alignment, and safety. Together, these give us a foundation for using AI responsibly in mental health settings.

But AI is advancing quickly. The next challenge is not only evaluating the tools we have today but also anticipating what will come tomorrow. The next chapter turns to the future—emerging trends, changing professional roles, and how psychologists can actively shape the direction of AI rather than simply adapt to it.

Framework Summary Table

To support practical comparison, the following table synthesizes key characteristics across the ten frameworks discussed. It is intended as a reference for readers seeking to select, adapt, or align evaluation approaches based on their specific context.

- **Framework Name** – The title of the framework or model.
- **Primary Focus** – The main evaluation goal (e.g., safety, integration, trust).
- **Domains Assessed** – The key categories or focus areas the framework evaluates, such as ethics (e.g., fairness, transparency), the clinical relevance of the tool (e.g., accuracy, applicability to therapy), the user experience (e.g., usability, accessibility), organizational fit (e.g., workflow integration, data governance), and technical performance (e.g., model validation, robustness).
- **Intended Users** – Who the framework is designed for (e.g., clinicians, developers, administrators).
- **Readiness Stage** – When the framework is most applicable (e.g., pre-deployment, in-practice, retrospective).

This summary offers a high-level map of the landscape, helping stakeholders quickly identify which frameworks best match their needs and where complementary tools may be used in combination.

AI Evaluation Frameworks Summary

Framework Name	Primary Focus	Domains Assessed	Intended Users	Readiness Stage
FAITA-Mental Health	Evaluating AI tools in mental health for clinical and ethical soundness	Credibility, User Experience, Agency, Equity, Transparency, Safety	Clinicians, Developers, Institutional Reviewers	Pre-deployment, In-practice
AI for IMPACTS	Evaluating real-world clinical impact and integration	Integration, Governance, Performance, Trust, Cost, Safety, Scalability	Healthcare Providers, Administrators, Regulators	In-practice, Retrospective
Thera-Turing Test (TTT)	Assessing therapeutic quality of conversational agents	Safety, Rapport, Fidelity, Cultural Sensitivity, Crisis Response	Clinicians, Developers, Researchers	Pre-deployment, In-practice
READI	Determining clinical readiness of AI tools for mental health	Safety, Privacy, Equity, Effectiveness, Engagement, Implementation	Clinicians, Administrators, Developers	Pre-deployment, In-practice
AIRI	Assessing organizational AI readiness in SMEs	Organizational, Governance, Business Value, Data, Infrastructure	Executives, Managers, Policy Advisors	Pre-deployment
AI-CAM	Evaluating organizational AI capability maturity	Business, Data, Technology, Organization, Skills, Ethics & Risk	Executives, Technologists, Strategy Teams	Pre-deployment, In-practice
AI Readiness Framework	Evaluating sociotechnical readiness for AI adoption	Technologies, Activities, Boundaries, Goals	Executives, Transformation Leads, IT Strategists	Pre-deployment
ISO/IEC JTC 1/SC 42	Standards-based AI governance and lifecycle management	Foundational, Data, Risk, Lifecycle, Ethics	Technology Providers, Regulators, Compliance Teams	All stages
Responsible AI Question Bank	Operationalizing ethical AI principles	Human Agency, Robustness, Privacy, Transparency, Fairness, Accountability	Executives, Risk Teams, Developers	Pre-deployment, In-practice
AI Governance and QA Framework	Governance and quality assurance in regulated industries	Culture, Skills, Governance Process, Data, Software, Infrastructure	Pharma Executives, QA Teams, Developers	All stages

Chapter 12: What's Next

Artificial intelligence (AI) is moving forward at an extraordinary pace, reshaping how people work, communicate, and make decisions across many fields. Psychology is no exception. While it is impossible to know exactly what the future will bring, especially in an area as complex as human behavior and mental health, we can still identify key trends and directions worth watching.

This chapter offers reflections on how the AI tools and applications covered in this book may develop further. The goal is not to predict with certainty, but to highlight ways these technologies could create even greater value for psychological practice in the years to come.

In-Session AI Co-Pilot Agents

In psychological care, an *In-Session AI Co-Pilot* refers to a digital assistant that works alongside clinicians as real-time support. Unlike autonomous systems that give direct recommendations to clients, a co-pilot stays in the background. Its role is to help the clinician by spotting subtle patterns, suggesting relevant interventions, and pulling together data during or between sessions. The clinician remains in full control, with the AI serving as an extra set of eyes and ears.

Earlier chapters (4 through 8) described many of the building blocks that make co-pilots possible. These include tools for documentation, triage, literature reviews, forensic risk modeling, and even limited therapeutic interactions. Each of these applications has been narrow, task-specific, and usually used outside the therapy room.

In-Session AI Co-Pilots bring these individual capabilities together in a more integrated form. Instead of simply analyzing information after the fact, they are expected to provide real-time insights, prompts, and decision support while therapy is taking place—always in a way that supports, rather than replaces, professional judgment.

For example, during a session, a co-pilot might detect changes in a client's speech patterns that suggest rising anxiety, or offer a prompt based on cognitive behavioral therapy (CBT) principles. After the session, it could help generate a summary, highlight recurring themes, or suggest follow-up questions for next time. In assessment or research contexts, the co-pilot could gather relevant studies, check for possible biases in interpretation, or generate alternative hypotheses from client data.

Why Humans Need In-Session AI Co-Pilots

Psychological practices often deal with high-stakes decisions where even small errors can have major consequences. Whether diagnosing conditions, conducting forensic risk assessments, or analyzing behavioral patterns, clinicians balance accuracy, efficiency, and judgment. AI co-pilots offer a way to ease this burden. They can act as collaborative partners, helping to reduce cognitive load, minimize errors, and free mental health professionals to focus more on client relationships and ethical decision-making.

By functioning as secondary reviewers and providing real-time insights, co-pilots may help clinicians make more accurate, data-informed choices while keeping professional judgment at the center. Over time, it is likely that in-session co-pilots will become more common, shifting from an occasional tool to a routine feature of practice.

In-Session AI Co-Pilots in Counseling and Clinical Psychology

In clinical settings, co-pilots could unify multiple capabilities into one integrated system. Rather than separate tools for documentation, progress monitoring, or diagnostics, these assistants may offer a holistic view by combining:

- **AI-driven symptom tracking:** Identifying early warning signs of mental health conditions before they worsen.

- **Speech and sentiment analysis:** Detecting subtle changes in tone or language that may signal rising anxiety, depression, or suicidal thinking.

- **Real-time alerts:** Prompting clinicians to reconsider assessments when AI identifies concerning shifts.

These features can support clinicians as "second reviewers," reinforcing clinical decisions while leaving therapists in full control.

In-Session AI Co-Pilots in Forensic and Risk Assessment Psychology

Forensic psychologists handle complex data from legal, medical, and psychological sources. AI co-pilots could strengthen this work by:

- **Analyzing case histories and legal precedents:** Spotting behavioral patterns linked to risk.

- **Comparing profiles with forensic databases:** Flagging individuals at high risk of harmful behavior.

- **Checking consistency in testimonies and reports:** Assisting experts in weighing evidence.

If widely adopted, co-pilots could help make forensic assessments more comprehensive, accurate, and fair, particularly in high-stakes criminal evaluations.

In-Session AI Co-Pilots in Psychological Research

In research, AI is already accelerating tasks like literature review and statistical analysis. Co-pilots may extend these abilities by:

- Summarizing and categorizing studies automatically.

- Identifying research gaps and generating new hypotheses.

- Detecting anomalies or patterns across large datasets.

Such capabilities would allow researchers to refine study designs more quickly, leading to faster and more reliable discoveries.

In-Session AI Co-Pilots in Industrial-Organizational (I-O) Psychology

In the workplace, co-pilots could support HR and leadership teams by:

- **AI-assisted hiring evaluations:** Analyzing traits and minimizing bias.

- **Employee engagement monitoring:** Detecting stressors before they lead to burnout.

- **Personalized leadership development:** Tailoring training based on AI-driven assessments.

Over time, AI co-pilots may become indispensable in supporting fair, data-based decisions that strengthen employee well-being and organizational health.

———————— ◗• ————————

AI Workflow Assistants

Earlier chapters, such as *Automating Administrative Tasks for Educators* and *Reducing Administrative Burden in Mental Health Practice*, highlighted how AI can help to ease the load of repetitive work. Tasks like data entry, scheduling, and documentation can be streamlined so that professionals spend more time on the heart of their work—caring for clients and students.

What is new in this section is the progression from basic task automation to broader workflow support. Instead of focusing only on isolated, repetitive duties, AI could connect multiple steps in a process. This means clinicians could have tools that not only file paperwork but also review documents, summarize findings, and provide support during sessions.

The shift is toward full workflow optimization. It is anticipated that AI will not only help to complete administrative chores, it will also analyze information, suggest actionable insights, and help coordinate tasks across a practice. The result could be a more integrated system that improves efficiency while supporting better care.

Processes That Would Benefit from an Integrated Workflow Solution

While today's AI tools are very good at handling single tasks like scheduling or documentation, many of the more complex workflows in psychological practice are still out of reach. These workflows require bringing together multiple types of information, adjusting as circumstances change, and sometimes making decisions in the moment. As AI evolves into more of a full workflow assistant, several areas could benefit:

Comprehensive Patient Case Management
Managing a patient's care often means pulling together records, therapy notes, assessments, and progress reports. At present, AI can review parts of this information, such as therapy notes or flagged concerns, but combining everything into one real-time view remains difficult. Future AI tools may be able to bring together data across electronic health records, therapy notes, and consultations, giving clinicians an integrated summary and tailored treatment recommendations.

Cross-Disciplinary Collaboration
Psychological care frequently involves multiple professionals—therapists, psychiatrists, social workers, and others. Right now, AI tools tend to work in silos, helping each professional with their own tasks. A more advanced AI system could share insights across disciplines, ensuring everyone works from the same dataset in real time and supporting more unified care.

Dynamic Treatment Plan Adjustments
Treatment plans often shift as patients improve or face new challenges. Today's AI tools can flag issues or make suggestions based on past data, but adjusting treatment dynamically in response to new observations, speech changes, or mood shifts requires more integration. Future systems could track these inputs continuously and recommend updates to keep plans timely, relevant, and personalized.

Predictive Workflow Management
AI already helps flag patients at risk or predict outcomes, but a more capable

workflow assistant could also streamline clinician operations. By monitoring patient loads, spotting scheduling conflicts, and anticipating staff needs, AI could automatically adjust calendars, redistribute tasks, and smooth out workflow.

Enhanced Client Engagement and Follow-Up

Simple reminders and follow-up systems exist today, but an integrated AI could adapt follow-up care more personally. For example, it might notice shifts in a patient's emotional state between sessions—using therapy notes, self-reports, or wearable data, and then recommend increasing session frequency, adding a check-in call, or providing resources.

When combined with an In-Session AI Co-Pilot that provides real-time insights during therapy, these workflow assistants could create a comprehensive support system. Together, they would reduce administrative load, strengthen collaboration, and give clinicians more time to focus on their patients.

Between-Session AI Therapy Assistants

We began this chapter with In-Session AI Co-Pilots, which support clinicians during therapy sessions by highlighting behavioral changes, suggesting possible interventions, and providing decision support in real time. We then looked at AI Workflow Assistants, which focus on reducing administrative burdens and coordinating practice operations, giving clinicians more time for direct patient care. Both tools are designed to help the clinician.

Between-Session AI Therapy Assistants extend this support to the patient directly. While co-pilots work alongside the clinician and workflow assistants streamline office tasks, these systems focus on empowering clients in their own mental health journey.

Drawing on tools such as voice recognition, facial expression analysis, and even wearable data, between-session assistants could provide support when patients are outside the therapy room. They could help guide relaxation practices, deliver CBT-based exercises, and even issue real-time alerts if signs of crisis emerge. The goal is personalized, ongoing care that adapts to the client's emotional and physical state.

Unlike clinician-facing tools, these therapy assistants would work directly with patients to build self-management skills, offer encouragement, and extend therapeutic support between sessions, helping to strengthen continuity of care and helping clients feel more supported day to day.

Evolving Beyond Text-Based Chatbots

Most AI therapy chatbots today communicate only through text. The next generation, however, is expected to be multi-modal, meaning they will be able to interpret not just words, but also voice, facial expressions, and data from wearable devices. This will make them much more responsive to a client's emotional and psychological state.

Some likely developments include:

- **Voice and Emotion Recognition**: AI tools will be able to better pick up on shifts in tone, pace, and speech patterns, helping detect signs of emotional distress or cognitive change. This could allow for real-time adjustments in how the system responds (Cohn et al., 2021).

- **Integration with Wearable Technology**: Devices like the Apple Watch or Oura Ring could feed data on stress indicators, such as heart rate or sleep disruptions, directly to an AI therapy assistant. In response, the assistant might recommend grounding exercises, relaxation practices, or even issue crisis alerts if needed (Torous et al., 2021).

- **Personalized Coping Strategies**: By learning from each client's history and progress, these assistants could provide coping tools that become more tailored over time, adjusting to the individual's evolving needs.

Proactive Mental Health Detection and Early Intervention

Between-session AI therapy assistants mark a shift from reactive treatment to proactive intervention. Instead of waiting until symptoms become severe, these systems could continuously analyze patterns in speech, text, and physical data to identify early warning signs of conditions such as anxiety, depression, or post-traumatic stress disorder (PTSD).

Some possible advancements include:

- **Speech and Language Analysis**: AI could examine tone, pauses, and sentence structure to detect subtle signs of distress, such as cognitive slowing or depressive speech patterns (Cohn et al., 2021).

- **Predictive Models for Suicide Prevention**: With client consent, AI could monitor text, social media activity, or voice interactions for red flags of suicidal ideation. If detected, the system could alert clinicians or designated contacts to enable faster intervention (Fitzpatrick et al., 2017).

- **Personalized Risk Scoring**: By integrating real-time data, AI could generate individualized risk scores, helping clinicians identify and prioritize patients who may be most in need of early support (Torous et al., 2021).

Real-Time Mood Tracking and Behavioral Modification

Between-session AI therapy assistants could become an important tool for long-term mental health management by continuously monitoring mood, habits, and emotional states. This ongoing tracking would give both clients and therapists the ability to adjust treatment plans in real time, based on up-to-date emotional and behavioral data.

Some possible features include:

- **Emotion and Behavior Analysis**: Using wearable devices and digital signals, AI could detect shifts in mood, energy levels, and stress through text, voice, and even facial microexpressions (Cohn et al., 2021).

- **Automated CBT and Behavioral Coaching**: AI systems might deliver personalized cognitive behavioral therapy (CBT) support—such as guided breathing, motivational prompts, or relaxation strategies—tailored to the client's current state (Fitzpatrick et al., 2017).

- **AI-Powered Virtual Exposure Therapy**: By combining Virtual Reality (VR) with AI, clients could engage in therapeutic scenarios like exposure therapy for phobias at their own pace, within a safe, controlled environment.

As these tools develop, they promise more proactive and personalized mental health care. Still, while they can support self-management and provide real-time interventions, there is also a need for approaches that encourage deeper relational growth and long-term change.

This is where AI-powered Relationship Care Platforms (RCPs) come in. By blending AI coaching, tools, content, and therapist collaboration, RCPs aim to help clients better understand relationship patterns and build healthier, more resilient connections over time.

AI Relationship Care Platforms

In Chapter 4, we looked at the rise of chatbots for relationship coaching. These tools are appealing because they are accessible and can provide quick support during stressful or emotionally charged moments. While helpful for immediate relief, chatbots often focus on isolated incidents. This makes them less effective in addressing the deeper, recurring patterns that usually bring people to seek relationship guidance. As a result, they may offer short-term comfort but rarely support lasting change.

AI Relationship Care Platforms (RCPs) represent a potential next step forward. Unlike basic chatbots, RCPs combine conversational support with ongoing tracking, assessment, and skill-building. The goal is not just to help in the moment but to help uncover and address the underlying dynamics that influence relationships over time.

RCPs also create a bridge between self-guided tools and professional therapy. Clients can choose to share their relationship data with their therapist, giving clinicians a clearer, more context-rich picture of patterns that may not surface in session. This allows therapy to move beyond isolated events and focus on the broader challenges shaping the relationship.

For example, companies such as *Relationship Workout* are developing systems that bring together:

- A **Relationship Fitness Assessment** that identifies individual relationship skill strengths and weaknesses, while informing the creation of a personalized Relationship Fitness plan.

- An **AI Relationship Coach** chatbot designed specifically for relationship coaching.

- **AI-powered tools** to help users assess, track, and strengthen their relationships.

- **Educational content** focused on building relationship skills and reducing unhelpful behaviors.

By combining these elements, RCPs provide clients with self-help tools that are grounded in real relationship patterns. They also give therapists the opportunity to incorporate this data into treatment, allowing for more personalized and effective guidance. Ultimately, this integrated approach aims to foster meaningful, lasting improvements in relationship health.

AI Relationship Coach Chatbot

AI Relationship Coach chatbots are already being used by people to manage stressful, emotionally charged situations. These tools provide on-demand, conversational support when someone isn't sure how to communicate, de-escalate conflict, or express their intentions. Their growing popularity reflects an important reality: many people look for relationship help before beginning therapy, or in cases where they may feel hesitant to take that step.

For these individuals, chatbots can act as an accessible, nonjudgmental bridge. They offer immediate support and can lower the barrier to eventually seeking therapy. Unlike mental health chatbots, which are usually designed for symptom screening or therapeutic exercises, relationship chatbots are focused on everyday interactions—helping with the wording of a difficult message, clarifying what someone wants to say, or finding ways to calm tension. They are not replacements for therapy and are not equipped to address deeper emotional struggles on their own.

One key limitation is that chatbot conversations usually stand alone. They can be helpful in the moment, but they do not connect insights across time, track recurring behavior, or identify long-term patterns. This makes it difficult to support lasting change. Users may find themselves returning repeatedly for immediate advice without getting help on the root causes of conflict.

AI-Powered Relationship Care Platforms (RCPs) represent a potential next step forward. Instead of leaving chatbots as isolated helpers, RCPs bring them into a more complete and holistic system of support. This includes structured journaling, behavior tracking, psychoeducation and skill-building resources, and—when appropriate—the ability to share selected data with a therapist. By linking real-time support with opportunities for reflection, growth, and professional guidance, RCPs help individuals move beyond reactive coping and toward long-term relationship health.

AI-Powered Relationship Journaling

A core feature of AI-Powered Relationship Care Platforms, alongside the AI Relationship Coach, will be a relationship journal. This tool allows users to capture experiences in real time—documenting not just what happened, but also how they felt and responded in the moment. Unlike traditional journaling, which often relies on memory and reflection after the fact, these platforms lower the barrier by using mobile-friendly options such as speech-to-text or by recording conversations with the AI Relationship Coach.

AI strengthens this process in two important ways. First, it can automatically summarize entries and identify recurring patterns, such as repeated conflicts, emotional triggers, or shifts in tone. Second, it can prompt deeper reflection by flagging emerging themes and offering guided questions or personalized feedback.

This journaling process supports self-awareness and also creates a valuable resource for therapy. Rather than arriving with vague recollections, users can share specific examples and long-term patterns with their therapist, providing a clearer and more actionable picture of their relationship dynamics.

Over time, the journal can track change—highlighting when stress levels are rising, when emotional withdrawal appears, or when communication is improving. Based on these insights, the platform can suggest targeted interventions, such as communication strategies, assertiveness exercises, or self-regulation practices.

While journaling has long been a therapeutic practice, AI-supported journaling brings structure, ongoing feedback, and continuity between self-reflection and professional care. In this way, it moves beyond being simply an outlet or record of past events. Instead, it becomes a tool for insight, behavior change, and sustained growth.

Role-Playing to Break Old Habits

One of the hardest parts of relationship growth is breaking old communication habits. Patterns like defensiveness, withdrawal, or escalation often play out automatically, making them difficult to notice in the moment and even harder to change. AI-powered role-playing tools within Relationship Care Platforms aim to help by creating a safe, low-pressure space to practice new strategies.

Through guided simulations, users could rehearse common challenges such as giving feedback, showing vulnerability, or navigating conflict. These exercises would mimic real conversations, giving people the chance to experiment with new approaches while receiving immediate, personalized feedback. For example, the AI might suggest alternative wording, a shift in tone, or better timing, helping users refine their communication in real time.

Importantly, the practice does not happen in isolation. The platform could track how often users role-play, how their responses change, and whether healthier patterns are emerging. This progress data can also be shared with a therapist, offering professionals richer insight into a client's growth and supporting more targeted guidance.

Beyond rehearsing scenarios, the AI could introduce evidence-based techniques like reframing negative thoughts or practicing active listening. By combining practice, feedback, and reinforcement, the platform helps people move from automatic reactions to intentional, skillful communication.

By turning insight into action, AI-powered role-playing could support more lasting behavior change. It could help clients build healthier ways of relating—inside and outside the therapy room.

While these more evolutionary tools discussed could create a more integrated approach to relationship support, the next stage is even more transformative: Client and Protocol Agents. These agents focus on securely managing data, streamlining collaboration, and tailoring mental health care across systems—representing a significant step forward in personalized, connected care.

Client and Protocol Agents

While earlier examples of AI in psychology represent gradual improvements, Client and Protocol Agents mark something much bigger: a new paradigm. Over the next several years, these intelligent systems could transform how people interact with mental health care—making support more personal, communication more seamless, and data protection much stronger. This shift moves us beyond today's fragmented systems, where records are often scattered and privacy concerns remain unresolved.

Client Agents could act as personal AI assistants for individuals. They would securely manage a person's mental health and wellness information, helping them organize, track, and share what matters most. Unlike current tools that primarily support clinicians, these agents would be designed to work directly with clients, giving them more control over their own care.

Protocol Agents, on the other hand, would represent therapists, clinics, or organizations. Their role would be to ensure that data shared between clients and professionals is handled in structured, ethical, and privacy-protected ways. By managing these exchanges, Protocol Agents would help reduce miscommunication, improve collaboration, and safeguard sensitive information.

Together, Client and Protocol Agents point to a future where individuals are more empowered in their care, and providers can focus on clinical insight instead of administrative barriers. This would represent not just an evolution, but a revolution in how mental health ecosystems may operate in a digital-first world.

What Are Client and Protocol Agents?

Client Agent: Think of this as a trusted digital companion that manages a person's health and psychological information. It securely stores records, analyzes patterns, and helps the individual decide what to share and with whom. Its main role is to protect privacy, give the client control, and support informed choices.

Protocol Agent: This is the counterpart on the professional side. A Protocol Agent represents a therapist, clinic, or organization. It ensures that when a Client Agent requests information or shares data, the exchange happens securely, ethically, and only with the patient's consent.

Together, these agents could change how mental health systems communicate. Instead of fragmented records and unclear boundaries, Client and Protocol Agents create a structured, privacy-first way of managing sensitive information.

In practice, a Client Agent could interact with Protocol Agents from therapists, forensic psychologists, mental health clinics, insurance companies, or even law enforcement—while still maintaining strict privacy safeguards to prevent unnecessary exposure of sensitive data.

Here are a few examples:

Matching Clients with the Right Psychologist

Finding the right therapist is one of the strongest predictors of positive outcomes in mental health care. AI-powered Client Agents could improve this process by:

- **Personalized matching:** Analyzing a person's communication style, therapy goals, and mental health history to suggest therapists who are the best fit.

- **Automated appointment scheduling:** Reducing frustration and delays by handling the logistics of scheduling quickly and seamlessly.

By improving the match between client and therapist, these systems could lower dropout rates, strengthen engagement, and foster more effective and personalized therapeutic experiences.

Secure and Selective Data Sharing for Improved Mental Health Treatment

Data privacy is one of the biggest concerns in mental health care. Many people are reluctant to share sensitive details with therapists because they worry about how their information might be used. AI-powered Client Agents could help address this challenge by acting as secure data managers. With these tools, individuals could:

- Share only the specific medical and psychological records needed by their therapist's Protocol Agent.
- Automatically transfer past therapy notes to support continuity of care.
- Connect wearable devices, such as heart rate or sleep trackers, to provide real-time insights into their mental health.
- Coordinate essential data across providers, including psychiatrists, physicians, and therapists, to create a more complete care plan.

This selective sharing approach would give individuals control over their personal information while helping clinicians access the right data at the right time. It reduces administrative effort and supports more accurate, holistic treatment.

Identifying Psychological Disorders with Physical Manifestations

AI-driven Client and Protocol Agents could strengthen the connection between psychological symptoms and physical health indicators, supporting a more holistic approach to care.

How these agents might help:

- **Monitoring physical health indicators:** A Client Agent could track stress markers, sleep cycles, and heart rate through wearable devices, offering insight into how physical health relates to mental well-being.

- **Cross-referencing biometric data:** The Client Agent could compare physical information with psychological symptoms reported during therapy sessions, highlighting trends or risks such as the early signs of a manic episode in bipolar disorder.

For example, the Client Agent might notice a pattern of disrupted sleep combined with changes in speech that point to a coming depressive episode. It could then alert the therapist's Protocol Agent, supporting early intervention before the situation escalates.

AI-Powered Personalized Mental Health Management

In the future, Client Agents may evolve into fully personalized mental health management systems. Possible features include:

- **Real-time mental health check-ins:** Individuals could receive ongoing AI-driven insights to help them track their mental health status.

- **Personalized coping strategies:** AI could suggest coping techniques based on past behaviors and therapy outcomes.

- **Continuous support:** The system could provide steady guidance without relying on intrusive monitoring, maintaining ethical boundaries.

These developments could encourage individuals to take a more active role in their mental health, supported by tailored strategies and tools for self-regulation.

Enhancing Privacy with AI-Facilitated Firewalls

A key advantage of AI-powered Client and Protocol Agents would be the potential to strengthen privacy and security. These systems could serve as protective firewalls between individuals and healthcare providers, making sure that:

- Only the data needed for treatment is shared with clinicians.

- Psychological records stay under the control of the individual.

- AI-based risk assessments are used responsibly and do not create unnecessary surveillance or discrimination.

This would mark a major change from current manual data-sharing practices, which often carry risks of privacy breaches, miscommunication, or gaps in mental health care.

Conclusion

In this chapter, we explored how AI may evolve in psychological practice, from co-pilots that support clinicians during sessions, to workflow assistants that reduce administrative burdens, to relationship care platforms that strengthen intimate relationships. What brings these innovations together is a shift away from single-purpose tools toward more integrated systems.

Looking ahead, the next step may be the development of fully connected platforms that combine today's separate tools into a seamless whole. These systems could streamline routine tasks while also generating deeper, data-informed insights. The result would be more time and energy for what matters most: delivering high-quality, personalized care. With careful design and ethical oversight, integrated AI systems could help make mental health care more accessible, more efficient, and more responsive to the needs of both clinicians and clients.

Chapter 13: A Few Last Comments

As we close this exploration of artificial intelligence in mental health care, AI holds real potential to help enhance the work of mental health professionals. By streamlining workflows, supporting diagnostic accuracy, and helping personalize care, AI can become a valuable partner—so long as its use is guided by ethics and careful implementation.

This chapter highlights ten key benefits AI may bring to mental health practice. These benefits reflect not only improvements in client care but also gains in efficiency for practitioners. Still, no matter how advanced these systems become, they cannot replace the human connection, judgment, and empathy that lie at the heart of psychology. Used responsibly, AI should strengthen and extend the expertise of mental health professionals, not take their place.

Alongside the benefits, this chapter also emphasizes the importance of transparency, privacy, and fairness in adopting AI. Psychologists must remain the final decision-makers, treating AI as a tool to support their work rather than an authority to follow uncritically. The future of AI in mental health will be shaped by how effectively governing bodies set standards and how thoughtfully clinicians integrate these technologies. Done well, AI can support therapeutic care while preserving the human relationship at its center.

Ten Core Benefits for AI in Psychological Practice

Below are ten consistent benefits of AI that we have observed during the writing of this book:

1. Greater Diagnostic and Predictive Accuracy

AI systems can spot patterns in behaviors, speech, and personal history that might be overlooked in a busy practice. In clinical psychology, predictive models may help flag early signs of distress, while in forensic psychology, they could strengthen risk assessments and evaluations. These tools are still developing, but with careful oversight, they hold promise for improving accuracy and confidence in decision-making.

2. Real-Time and Between-Session Monitoring

Instead of relying only on what happens in scheduled sessions, AI tools (such as wearables or digital check-ins) can offer continuous monitoring and feedback. This can help track progress, identify early warning signs of relapse, and support more proactive care. While this technology is still maturing, it points toward a future where support doesn't stop when the session ends.

3. A More Complete Picture of Client Well-being

AI can combine data from multiple sources—speech, text, physiological signals, and behavior—into a more holistic view of a client's mental health. By integrating these perspectives, clinicians may gain deeper insights into what's happening in a client's life, though validation is still needed to ensure accuracy and clinical usefulness.

4. Enhanced Personalization of Treatment Plans

Because AI can track long-term patterns, it may help tailor treatment plans more closely to each client's needs and preferences. This could improve engagement and outcomes by suggesting interventions that feel more relevant to the individual. As with all AI tools, these approaches will require careful validation across diverse populations.

5. Improved Accessibility to Psychological Support

AI-powered chatbots and virtual assistants can offer immediate, low-cost guidance, making support more accessible in communities with limited mental health resources. While these tools aren't a replacement for therapy, they can provide helpful, timely support when human clinicians are unavailable.

6. Bias Mitigation (When Built Responsibly)

If designed with diverse, high-quality data, AI has the potential to reduce certain biases in assessments and decision-making. This could promote more equitable care by supporting consistency across populations. That said, vigilance and oversight are essential to ensure AI does not introduce new forms of bias.

7. Enhanced Consistency

AI systems apply the same criteria each time, helping reduce variability caused by human interpretation. This can make assessments and interventions more uniform and reliable, especially in structured decision-making contexts. Still, clinicians remain essential for ensuring that "consistent" also means clinically appropriate.

8. Improved Efficiency

By taking on administrative tasks like documentation, triage, or data processing, AI can save time and free clinicians to focus on direct client care. This is one of the most immediate benefits already being seen in practice, though alignment with clinical standards is critical to avoid errors or inefficiencies.

9. Improved Engagement

AI can help clients stay connected between sessions through reminders, check-ins, and personalized resources. These tools may increase motivation, provide continuity of care, and make clients feel more supported in their growth journey. Done thoughtfully, this can strengthen the therapeutic process rather than distract from it.

10. Augmentation of, Not Replacement for, Professional Judgment

Perhaps the most important benefit: AI is designed to support, not replace, the expertise of psychologists. It can highlight patterns or offer suggestions, but the clinician remains the ultimate decision-maker, applying professional training, ethical judgment, and personal understanding of each client's context.

Five Guiding Principles for Responsible AI Use

To help ensure that AI tools truly benefit both clients and professionals, their use must be guided by a few core principles. These principles help keep AI trustworthy, transparent, and equitable across diverse clinical settings:

1. Human Oversight is Essential

AI should serve as an aid—not a substitute—for professional judgment. Especially in high-stakes contexts, clinicians must interpret AI outputs, apply them within the client's unique context, and make the final decisions. Ethical care requires that expertise, empathy, and clinical judgment remain at the center.

2. Transparency is Critical

For AI to be responsibly integrated, clinicians, clients, and organizations need clear, understandable information about how systems function. This includes knowing what data is used, how outputs are generated, and where limitations exist. Transparency fosters accountability and helps clients feel safe engaging with AI-supported care.

3. Cultural and Contextual Sensitivity Must Be Prioritized

AI systems should reflect the diversity of the populations they serve. Tools trained only on narrow or homogeneous data risk misclassifying behaviors and overlooking important cultural factors. Ensuring cultural and contextual sensitivity is essential to delivering fair, relevant, and respectful care.

4. Privacy and Confidentiality Must Be Safeguarded

AI must meet the highest standards for data security and confidentiality. This includes compliance with regulations such as HIPAA and the commitment to protect sensitive client information. Safeguarding privacy is not just a legal requirement—it is central to maintaining the therapeutic alliance.

5. Equity Must Be Continuously Monitored

Bias in AI can undermine trust and harm clients. Systems should be regularly reviewed and audited to detect and correct unfair patterns in how they process or apply information. Continuous monitoring ensures that AI contributes to equitable care and supports positive outcomes across all client groups.

Call to Action

As we look ahead, the opportunity for those working in psychological practice is not only to adopt AI tools but to actively shape how these technologies are integrated into care. This requires building AI literacy to critically evaluate tools, advocating for ethical standards, and engaging in interdisciplinary collaboration to ensure AI reflects the core values of psychology. Ethical considerations must remain central, as mental health professionals must do their part to safeguard transparency, equity, and client well-being at every stage.

The promise of AI to transform therapeutic settings is immense, but it is vital that psychologists maintain oversight, ensuring that technology functions as a supportive ally rather than a substitute for human insight, empathy, and ethical judgment.

Now, more than ever, psychologists must take proactive steps to:

- **Educate** themselves and future generations about AI and its ethical implications.

- **Advocate** for responsible AI development and integration that prioritizes patient-centered care.

- **Participate** in shaping AI's role across research, clinical practice, and mental health support, ensuring it strengthens the human connection at the core of effective therapy.

The future of AI in psychology holds tremendous potential. But it is up to professionals to ensure its use remains responsible, ethical, and human-centered. By embracing innovation while staying anchored in professional ethics, mental health professionals can guide AI to become a true enabler of more accessible, personalized, and compassionate care, helping to build a brighter future for mental health worldwide.

References

Abedi, A., Colella, T. J. F., Pakosh, M., & Khan, S. S. (2024). Artificial intelligence-driven virtual rehabilitation for people living in the community: A scoping review. *NPJ Digital Medicine*, 7(25). https://doi.org/10.1038/s41746-024-00998-w

Agrawal, R., Monteiro, K., Narasimhamurti, N., Sharma, S., Suryawanshi, A., Bariya, A., Narvekar, S., Bagnoli, L., Saxena, M., Magoun, L., Parsekar, S., Pozuelo, J., Lesh, N., Sood, M., Sharma, T., Yadav, H., Bhan, A., Nadkarni, A., & Patel, V. (2025). A conversational agent (PracticePal) to support the delivery of a brief behavioral activation treatment for depression in rural India: Development and pilot-testing study. JMIR FORMATIVE RESEARCH, 9, e73563. https://doi.org/10.2196/73563

Akpobome, Omena. (2024). The Impact of Emerging Technologies on Legal Frameworks: A Model for Adaptive Regulation. *International Journal of Research Publication and Reviews*. 5. 5046-5060. 10.55248/gengpi.5.1024.3012.

Altrabsheh, E., Heitmann, M., & Lochbronner, A. (2022). *AI Governance and QA Framework: AI Governance Process Design*. Pharmaceutical Engineering, July/August 2022. https://ispe.org/pharmaceutical-engineering/july-august-2022/ai-governance-and-qa-framework-ai-governance-process

American Psychological Association (APA) Task Force on AI Ethics in Psychology. (2024). *Guidelines for responsible AI implementation in psychological practice. American Psychological Association.*

Angwin, J., Larson, J., Mattu, S., & Kirchner, L. (2016). Machine bias: There's software used across the country to predict future criminals. And it is biased against Black defendants. *ProPublica*. https://www.propublica.org/article/machine-bias-risk-assessments-in-criminal-sentencing

Asfahani, A. M. (2022). The impact of artificial intelligence on industrial-organizational psychology: A systematic review. *The Journal of Behavioral Science*, 17(3), 125–139. https://www.researchgate.net/publication/364421222

Babu, A., & Joseph, A. P. (2024). Artificial intelligence in mental healthcare: Transformative potential vs. the necessity of human interaction. *Frontiers in Psychology*, 15, 1378904. https://doi.org/10.3389/fpsyg.2024.1378904

Baines, J. I., Dalal, R. S., Ponce, L. P., & Tsai, H.-C. (2024). Advice from artificial intelligence: A review and practical implications. *Frontiers in Psychology*, 15, 1390182. https://doi.org/10.3389/fpsyg.2024.1390182

Bankins, S., Ocampo, A. C., Marrone, M., Restubog, S. L. D., & Woo, S. E. (2023). A multilevel review of artificial intelligence in organizations: Implications for organizational behavior research and practice. *Journal of Organizational Behavior*, 45(2), 159–182. https://doi.org/10.1002/job.2735

Balcombe, L. (2023). AI Chatbots in Digital Mental Health. *Informatics, 10*(4), 82. https://doi.org/10.3390/informatics10040082

Banos, O., Comas-González, Z., Medina, J., Polo-Rodríguez, A., Gil, D., Peral, J., Amador, S., & Villalonga, C. (2024). Sensing technologies and machine learning methods for emotion recognition in autism: Systematic review. International Journal of Medical Informatics, 187, 105469. https://doi.org/10.1016/j.ijmedinf.2024.105469

Bedi, G., Carrillo, F., Cecchi, G. A., Slezak, D. F., Sigman, M., Mota, N. B., Ribeiro, S., Javitt, D. C., Copelli, M., & Corcoran, C. M. (2015). Automated analysis of free speech predicts psychosis onset in high-risk youths. *NPJ Schizophrenia*, 1, 15030. https://doi.org/10.1038/npjschz.2015.30

Beg, M. J., Verma, M., Chanthar, K. M. M. V., & Verma, M. K. (2024). Artificial intelligence for psychotherapy: A review of the current state and future directions. *Indian Journal of Psychological Medicine*. Advance online publication. https://doi.org/10.1177/02537176241260819

Banerjee, D., Islam, K., Xue, K., & others. (2019). A deep transfer learning approach for improved post-traumatic stress disorder diagnosis. *Knowledge and Information Systems*, 60, 1693–1724. https://doi.org/10.1007/s10115-019-01337-2

Basei de Paula, P., Bruneti Severino, J., Berger, M., Veiga, M., Parente Ribeiro, K., Loures, F., Todeschini, S., Roeder, E., & Marques, G. (2025). Improving documentation quality and patient interaction with AI: a tool for transforming medical records—an experience report. *Journal Of Medical Artificial Intelligence, 8*. doi:10.21037/jmai-24-213

Bentley, S. V. (2025). Knowing you know nothing in the age of generative AI. *Humanities and Social Sciences Communications*, 12, 409. https://doi.org/10.1057/s41599-025-04731-0

Benrimoh, D., Tanguay-Sela, M., Perlman, K., Israel, S., Mehltretter, J., Armstrong, C., Fratila, R., Parikh, S. V., Karp, J. F., Heller, K., Vahia, I. V., Blumberger, D. M., Karama, S., Vigod, S. N., Myhr, G., Martins, R., Rollins, C., Popescu, C., Lundrigan, E., Snook, E., Wakid, M., Williams, J., Soufi, G., Perez, T., Tunteng, J., Rosenfeld, K., Miresco, M., Turecki, G., Cardona, L. G., Linnaranta, O., & Margolese, H. C. (2021). Using a simulation centre to evaluate preliminary acceptability and impact of an artificial intelligence-powered clinical decision support system for depression treatment on the physician–patient interaction. *BJPsych Open, 7*(1), e22. https://doi.org/10.1192/bjo.2020.127

Binns, R. (2018). Fairness in machine learning: Lessons from political philosophy. *Proceedings of the 2018 Conference on Fairness, Accountability, and Transparency (FAT),* 149–159. https://doi.org/10.48550/arXiv.1712.03586

Biofourmis. (2022, February 9). Biofourmis launches virtual chronic condition management and monitoring. *MobiHealthNews.* https://www.mobihealthnews.com/news/biofourmis-launches-virtual-chronic-condition-management-monitoring

Brundage, M., Avin, S., Clark, J., Toner, H., Eckersley, P., Garfinkel, B., … & Amodei, D. (2020). *Toward trustworthy AI development: Mechanisms for supporting verifiable claims* (arXiv preprint arXiv:2004.07213). https://arxiv.org/abs/2004.07213

Byrnes, J., & Robinson, M. (2024). Transparency and authority concerns with using AI to make ethical recommendations in clinical settings. *Nursing Ethics,* 0(0). https://doi.org/10.1177/09697330241307317

Budiraharjo, M. H., Muhammad, A. H., & Kusnawi. (2024). MSME AI readiness analysis using the AIRI framework. *Jurnal Teknologi dan Informasi Universitas Lambung Mangkurat (JTIULM),* 9(2), 33–44. https://jtiulm.ti.ft.ulm.ac.id/index.php/jtiulm/article/view/307

Bunge, E. L., & Desage, C. (2025). A framework for evaluating mental health artificial intelligence-based conversational agents. *Journal of Technology in Behavioral Science. Advance online publication.* https://doi.org/10.1007/s41347-025-00519-w

Büscher, R., Winkler, T., Mocellin, J., & others. (2024). A systematic review on passive sensing for the prediction of suicidal thoughts and behaviors. *NPJ Mental Health Research,* 3, 42. https://doi.org/10.1038/s44184-024-00089-4

Butler, T., Espinoza-Limón, A., & Seppälä, S. (2021). *Towards a Capability Assessment Model for the Comprehension and Adoption of AI in Organisations*. Journal of AI, Robotics & Workplace Automation, 1(1), 18–33. https://doi.org/10.48550/arXiv.2305.15922

Cai, C. J., Winter, S., Steiner, D., Wilcox, L., & Terry, M. (2019). "Hello AI": Uncovering the onboarding needs of medical practitioners for human–AI collaborative decision-making. *Proceedings of the ACM on Human-Computer Interaction*, 3(CSCW), 1–24. https://doi.org/10.1145/3359206

Chang, B. P., & Srivastava, S. (2024). AI and the future of psychological science: A data-centric perspective (arXiv preprint arXiv:2402.14424). https://doi.org/10.48550/arXiv.2402.14424

Chang, X. (2023). Gender bias in hiring: An analysis of the impact of Amazon's recruiting algorithm. Proceedings of the 2023 International Conference on Management Research and Economic Development. https://doi.org/10.54254/2754-1169/23/20230367

Chouldechova, A., & G'Sell, M. (2017). Fairer and more accurate, but for whom? *arXiv Preprint*, arXiv:1707.00046. https://arxiv.org/abs/1707.00046

Cohen, I. G. (2020). Informed consent and medical artificial intelligence: What to tell the patient? *Georgetown Law Journal*, 108(6), 1425–1469. https://www.law.georgetown.edu/georgetown-law-journal/in-print/volume-108/volume-108-issue-6-june-2020/informed-consent-and-medical-artificial-intelligence-what-to-tell-the-patient/

Cohn, J. F., Kruez, T. S., Matthews, I., Yang, Y., Nguyen, H., Padilla, M. T., Zhou, F., & De la Torre, F. (2019). Detecting depression from facial actions and vocal prosody. *Proceedings of the IEEE International Conference on Automatic Face & Gesture Recognition*, 1-7. https://www3.cs.stonybrook.edu/~minhhoai/papers/acii-paper_final.pdf

Coley, R. Y., Johnson, E., Simon, G. E., Cruz, M., & Shortreed, S. M. (2021). Racial/ethnic disparities in the performance of prediction models for death by suicide after mental health visits. *JAMA Psychiatry, 78*(7), 726–734. https://doi.org/10.1001/jamapsychiatry.2021.0493

Council on Criminal Justice. (2025). *Department of Justice report on AI in criminal justice: Key takeaways*. Council on Criminal Justice. Retrieved from https://counciloncj.org/doj-report-on-ai-in-criminal-justice-key-takeaways/

Crompton, H., & Burke, D. (2023). Artificial intelligence in higher education: The state of the field. *International Journal of Educational Technology in Higher Education*, 20(22). https://doi.org/10.1186/s41239-023-00392-8

Cross, J. L., Choma, M. A., & Onofrey, J. A. (2024). Bias in medical AI: Implications for clinical decision-making. *PLOS Digital Health*, 3(11), e0000651. https://doi.org/10.1371/journal.pdig.0000651

Cruz-Gonzalez, P., He, A. W.-J., Lam, E. P., et al. (2025). Artificial intelligence in mental health care: A systematic review of diagnosis, monitoring, and intervention applications. *Psychological Medicine, 55*, e18. https://doi.org/10.1017/S0033291724003295

Day, A., Woldgabreal, Y., & Butcher, L. (2022). Cultural bias in forensic assessment: Considerations and suggestions. *In Forensic Psychology and Criminology* (pp. Chapter 16). https://doi.org/10.4324/9781003230977-16

De Choudhury, M., Gamon, M., Counts, S., & Horvitz, E. (2021). Predicting Depression via Social Media. *Proceedings of the International AAAI Conference on Web and Social Media*, 7(1), 128-137. https://doi.org/10.1609/icwsm.v7i1.14432

Deka, C., Shrivastava, A., Abraham, A. K., Nautiyal, S., & Chauhan, P. (2024). *AI-based automated speech therapy tools for persons with speech sound disorders: A systematic literature review*. Research Square. https://doi.org/10.21203/rs.3.rs-1517404/v2

DeMatteo, D., & Heilbrun, K. (2023). Implications for artificial intelligence in forensic psychological practice and board certification. *American Board of Professional Psychology*. https://abpp.org/newsletter-post/implications-for-artificial-intelligence-in-forensic-psychological-practice-and-board-certification/

Desai, A., Abdelhamid, M., & Padalkar, N. R. (2024). What is reproducibility in artificial intelligence and machine learning research? *arXiv Preprint*, arXiv:2407.10239. https://doi.org/10.48550/arXiv.2407.10239

Doshi-Velez, F., & Kim, B. (2017). Towards a rigorous science of interpretable machine learning. *arXiv preprint*. https://doi.org/10.48550/arXiv.1702.08608

Dressel, J., & Farid, H. (2018). The accuracy, fairness, and limits of predicting recidivism. *Science Advances*, 4(1), eaao5580. https://doi.org/10.1126/sciadv.aao5580

Dwianto, A., Kusuma, S., & Junengsih. (2024). Artificial intelligence in performance evaluation (Case study of PT. Pos Indonesia employees). *bit-Tech*, 7, 348–356. https://doi.org/10.32877/bt.v7i2.1817

Dwyer, D. B., Falkai, P., & Koutsouleris, N. (2018). Machine learning approaches for clinical psychology and psychiatry. *Annual Review of Clinical Psychology*, 14, 91-118. https://doi.org/10.1146/annurev-clinpsy-032816-045037

European Parliament & Council of the European Union. (2024). *Regulation (EU) 2024/1689 of the European Parliament and of the Council of 13 June 2024 laying down harmonised rules on artificial intelligence and amending certain Union legislative acts (Artificial Intelligence Act). Official Journal of the European Union*, L 202, 12 July 2024, 1–144. https://eur-lex.europa.eu/eli/reg/2024/1689/oj

Ferguson, B. J., Hamlin, T., Lantz, J. F., Villavicencio, T., Coles, J., & Beversdorf, D. Q. (2019). Examining the association between electrodermal activity and problem behavior in severe autism spectrum disorder: A feasibility study. *Frontiers in Psychiatry*, 10, 654. https://doi.org/10.3389/fpsyt.2019.00654

Fitzpatrick, K. K., Darcy, A., & Vierhile, M. (2017). Delivering cognitive behavior therapy to young adults with symptoms of depression and anxiety using a fully automated conversational agent (Woebot): A randomized controlled trial. *JMIR Mental Health*, 4(2), e19. https://doi.org/10.2196/mental.7785

Flodén, J. (2024). AI-generated student grades and teacher perceptions – A case study. *Education and Information Technologies*. https://doi.org/10.1007/s10639-024-12137-w

Flodén, J. (2025). Grading exams using large language models: A comparison between human and AI grading of exams in higher education using ChatGPT. *British Educational Research Journal*, 51, 201–224. https://doi.org/10.1002/berj.4069

Floridi, L., Cowls, J., Beltrametti, M., Chatila, R., Chazerand, P., Dignum, V., … & Vayena, E. (2018). AI4People—An ethical framework for a good AI society: Opportunities, risks, principles, and recommendations. *Minds and Machines*, 28(4), 689–707. https://doi.org/10.1007/s11023-018-9482-5

Fraser, K. C., Meltzer, J. A., & Rudzicz, F. (2016). Linguistic features identify Alzheimer's disease in narrative speech. *Journal of Alzheimer's Disease*, 49(2), 407–422. https://doi.org/10.3233/JAD-150520

Freeman, D., Reeve, S., Robinson, A., Ehlers, A., Clark, D., Spanlang, B., & Slater, M. (2017). Virtual reality in the assessment, understanding, and treatment of mental health disorders. *Psychological Medicine*, 47(14), 2393–2400. https://doi.org/10.1017/S003329171700040X

Fulmer, R. (2019). Artificial intelligence and counseling: Four levels of implementation. *Theory & Psychology*, 29(3), 318–338. https://doi.org/10.1177/0959354319853045

Genomind. (2021, May 27). *Medicare provides coverage of the Genomind PGx test*. https://genomind.com/providers/medicare-provides-coverage-of-the-genomind-professional-pgx-express-test/

Glenn, J. M., Bryk, K., Myers, J. R., Anderson, J., Onguchi, K., McFarlane, J., & Ozaki, S. (2023). The efficacy and practicality of the Neurotrack Cognitive Battery assessment for utilization in clinical settings for the identification of cognitive decline in an older Japanese population. *Frontiers in Aging Neuroscience*, 15, 1206481. https://doi.org/10.3389/fnagi.2023.1206481.

Golden, A., & Aboujaoude, E. (2024). Describing the framework for AI tool assessment in mental health and applying it to a generative AI obsessive-compulsive disorder platform: Tutorial. *JMIR Formative Research*, 8, e62963. https://doi.org/10.2196/62963

Gordon, F. (2019). [Review of the book *Automating inequality: How high-tech tools profile, police, and punish the poor,* by V. Eubanks]. *Law, Technology and Humans*, 1, 162–164. https://doi.org/10.5204/lthj.v1i0.1386

Habicht, J., Dina, L.-M., McFadyen, J., Stylianou, M., Harper, R., Hauser, T. U., & Rollwage, M. (2025). Generative AI–enabled therapy support tool for improved clinical outcomes and patient engagement in group therapy: Real-world observational study. JOURNAL OF MEDICAL INTERNET RESEARCH, 27, e60435. https://doi.org/10.2196/60435

Hagendorff, T. The Ethics of AI Ethics: An Evaluation of Guidelines. *Minds & Machines* 30, 99–120 (2020). https://doi.org/10.1007/s11023-020-09517-8

Hagendorff, T. (2024). Deception abilities emerged in large language models. *Proceedings of the National Academy of Sciences of the United States of America*, 121(11), e2317967121. https://doi.org/10.1073/pnas.2317967121

Hagendorff, T., Dasgupta, I., Binz, M., Chan, S. C. Y., Lampinen, A., Wang, J. X., Akata, Z., & Schulz, E. (2024). Machine psychology: Understanding AI behavioral patterns. *Google DeepMind Research, 6(2)*, 102-125. https://doi.org/10.1007/s54321-024-00234-9

Haque, A., Guo, M., Miner, A. S., & Fei-Fei, L. (2018). Measuring depression symptom severity from spoken language and 3D facial expressions. In *NeurIPS Machine Learning for Health (*ML4H) Workshop. https://arxiv.org/abs/1811.08592

Haque, M. D. R., & Rubya, S. (2023). An overview of chatbot-based mobile mental health apps: Insights from app description and user reviews. *JMIR mHealth and uHealth*, 11, e44838. https://doi.org/10.2196/44838

Harley, J., Chan, J., Darnell, D., et al. (2024). Randomized Controlled Trial of a Generative AI Chatbot for Mental Health Support. *NEJM AI*. https://doi.org/10.1056/Aloa2400802

Hogan, N. R., Davidge, E. Q., & Corabian, G. (2021). *On the ethics and practicalities of artificial intelligence, risk assessment, and race. Journal of the American Academy of Psychiatry and the Law*, 49(3), 326–334. https://doi.org/10.29158/JAAPL.200116-20

Health Resources and Services Administration. (2023). *Behavioral health workforce: HRSA programs and policy.* U.S. Department of Health and Human Services. https://bhw.hrsa.gov/sites/default/files/bureau-health-workforce/Behavioral-Health-Workforce-Brief-2023.pdf

Holmström, J. (2022). From AI to digital transformation: The AI readiness framework. *Business Horizons*, 65(3), 329–339. https://doi.org/10.1016/j.bushor.2021.03.006

Huckvale, K., Venkatesh, S., & Christensen, H. (2019). Toward clinical digital phenotyping: A timely opportunity to consider purpose, quality, and safety. *npj Digital Medicine*, 2, Article 88. https://doi.org/10.1038/s41746-019-0166-1

International Organization for Standardization / International Electrotechnical Commission. (2023). *ISO/IEC JTC 1/SC 42 Artificial Intelligence: Standards and Projects Overview.* Retrieved from https://www.iso.org/committee/6794475.html

Jiang, F., Jiang, Y., Zhi, H., Dong, Y., Li, H., Ma, S., Wang, Y., Dong, Q., Shen, H., & Wang, Y. (2017). Artificial intelligence in healthcare: Past, present and future. *Stroke and Vascular Neurology*, 2(4), 230–243. https://doi.org/10.1136/svn-2017-000101

Jacob, C., Brasier, N., Laurenzi, E., Heuss, S., Mougiakakou, S.-G., Cöltekin, A., & Peter, M. K. (2025). *AI for IMPACTS framework for evaluating the long-term real-world impacts of AI-powered clinician tools: Systematic review and narrative synthesis.* Journal of Medical Internet Research, 27, e67485. https://doi.org/10.2196/67485

Jobin, Anna & Ienca, Marcello & Vayena, Effy. (2019). The global landscape of AI ethics guidelines. *Nature Machine Intelligence.* 1. 10.1038/s42256-019-0088-2.

Kiyasseh, D., Cohen, A., Jiang, C., & Altieri, N. (2024). A framework for evaluating clinical artificial intelligence systems without ground-truth annotations. *Nature Communications*, 15, 1808. https://doi.org/10.1038/s41467-024-46000-9

Klimova, B., & Pikhart, M. (2025). Exploring the effects of artificial intelligence on student and academic well-being in higher education: A mini-review. *Frontiers in Psychology*, 16, Article 1498132. https://doi.org/10.3389/fpsyg.2025.1498132

Kiron, D., Spindel, B., Unruh, G., & Hancock, B. (2022). *Workforce ecosystems: A new strategic approach to the future of work.* MIT Sloan Management Review. https://sloanreview.mit.edu/projects/workforce-ecosystems-a-new-strategic-approach-to-the-future-of-work/

Klimova, B., & Pikhart, M. (2025). Exploring the effects of artificial intelligence on student and academic well-being in higher education: A mini-review. *Frontiers in Psychology*, 16, Article 1498132. https://doi.org/10.3389/fpsyg.2025.1498132

Ko, J., & Li, L. (2021). The application of deep learning in mental health assessment: A review. *Journal of Affective Disorders*, 286, 1-12.

Kollins, S. H., DeLoss, D. J., Cañadas, E., Lutz, J., Findling, R. L., Keefe, R. S. E., Epstein, J. N., Cutler, A. J., & Faraone, S. V. (2020). A novel digital intervention for actively reducing severity of paediatric ADHD (STARS-ADHD): A randomised controlled trial. *The Lancet Digital Health*, 2(4), e168–e178. https://doi.org/10.1016/S2589-7500(20)30017-0

Kraljevic, Z., Bean, D., Shek, A., Bendayan, R., Hemingway, H., Yeung, J. A., Deng, A., Balston, A., Ross, J., Idowu, E., Teo, J. T., & Dobson, R. J. B. (2024). *Foresight—a generative pretrained transformer for modelling of patient timelines using electronic health records: a retrospective modelling study.* The Lancet Digital Health, 6, e281–e290. https://doi.org/10.1016/S2589-7500(24)00025-6

Kuhail, I., Boese, M., El-Assady, M., & El-Gayar, M. (2024). *Human-human vs human-AI therapy: An empirical study. Proceedings of the 2024 CHI Conference on Human Factors in Computing* Systems. https://doi.org/10.1145/3613904.3642105

Lai, W. Y. W., & Lee, J. S. (2024). A systematic review of conversational AI tools in ELT: Publication trends, tools, research methods, learning outcomes, and antecedents. *Computers and Education: Artificial Intelligence*, 7, 100291. https://doi.org/10.1016/j.caeai.2024.100291

Lee, E. E., Torous, J., De Choudhury, M., Depp, C. A., Graham, S. A., Kim, H. C., Paulus, M. P., Krystal, J. H., & Jeste, D. V. (2021). Artificial intelligence for mental health care: Clinical applications, barriers, facilitators, and artificial wisdom. *Biological Psychiatry: Cognitive Neuroscience and Neuroimaging*, 6(9), 856–864. https://doi.org/10.1016/j.bpsc.2021.02.001

Lee, S. U., Mantel, M., Abhishek, A., Jain, N., & Whittlestone, J. (2024). *Responsible AI Question Bank: A Tool to Support AI Risk Assessment and Governance.* arXiv preprint arXiv:2408.11820. https://doi.org/10.48550/arXiv.2408.11820

Li, H., Zhang, R., Lee, YC. *et al.* Systematic review and meta-analysis of AI-based conversational agents for promoting mental health and well-being. *npj Digit. Med.* **6**, 236 (2023). https://doi.org/10.1038/s41746-023-00979-5

Lin, B., Cecchi, G., & Bouneffouf, D. (2023). Psychotherapy AI companion with reinforcement learning recommendations and interpretable policy dynamics. In *Companion Proceedings of the ACM Web Conference 2023 (WWW '23 Companion)* (pp. 756–759). ACM. https://doi.org/10.1145/3543873.3587623

Lin, H., & Chen, Q. (2024). Artificial intelligence (AI)-integrated educational applications and college students' creativity and academic emotions: Students and teachers' perceptions and attitudes. *BMC Psychology*, 12(487). https://doi.org/10.1186/s40359-024-01979-0

Lipton, Z. C. (2016). *The mythos of model interpretability.* arXiv. https://doi.org/10.48550/arXiv.1606.03490

Liu, T., Zhao, H., Liu, Y., Wang, X., & Peng, Z. (2024). ComPeer: A generative conversational agent for proactive peer support. ARXIV PREPRINT ARXIV:2407.18064. https://doi.org/10.48550/arXiv.2407.18064

Lockwood, A., & Brown, J. (2024). *Mitigating AI bias in school psychology: Toward equitable and ethical implementation.* PsyArXiv. https://doi.org/10.31234/osf.io/mh4rj

Luo, Y., & Liu, Q. (2023). *Cybersecurity and privacy in digital mental health: concerns, challenges, and recommendations. Frontiers in Digital Health*, 5, Article 1242264. https://doi.org/10.3389/fdgth.2023.1242264

Luoma, A., Cheng, K., Srivastava, A., & Choudhury, M. (2024). *Exploring the frontiers of LLMs in psychological applications: A comprehensive review* (arXiv:2401.01519). arXiv. https://arxiv.org/abs/2401.01519

Luxton, D. D. (2016). *An introduction to artificial intelligence in behavioral and mental health care*. Academic Press.

Manole, A., Cârciumaru, R., Brînzaș, R., & Manole, F. (2024). Harnessing AI in Anxiety Management: A Chatbot-Based Intervention for Personalized Mental Health Support. *Information*, *15*(12), 768. https://doi.org/10.3390/info15120768

Malouin-Lachance, A., Capolupo, J., Laplante, C., & Hudon, A. (2025). Does the digital therapeutic alliance exist? Integrative review. *JMIR Mental Health*, 12, e69294. https://doi.org/10.2196/69294

Matochová, Jana & Kowaliková, Petra. (2024). Transforming Higher Education: Psychological and Sociological Perspective (the use of artificial intelligence). *R&E-SOURCE*. 176-181. 10.53349/resource.2024.is1.a1253.

Marler, J. D., Fujii, C. A., Wong, K. S., Galanko, J. A., Balbierz, D. J., & Utley, D. S. (2020). Assessment of a personal interactive carbon monoxide breath sensor in people who smoke cigarettes: Single-arm cohort study. *Journal of Medical Internet Research*, 22(10), e22811. https://doi.org/10.2196/22811

Marmar, C. R., Brown, A. D., Qian, M., Laska, E., Siegel, C., Li, M., Abu-Amara, D., Tsiartas, A., Richey, C., Smith, J., Knoth, B., & Vergyri, D. (2019). Speech-based markers for posttraumatic stress disorder in US veterans. *Depression and Anxiety*, 36(7), 607–616. https://doi.org/10.1002/da.22890

Mansoor, M. A., & Ansari, K. H. (2024). Early detection of mental health crises through artificial-intelligence-powered social media analysis: A prospective observational study. *Journal of Personalized Medicine, 14*(958). https://doi.org/10.3390/jpm14090958

Marr, B. (2019, December 14). The amazing ways how Unilever uses artificial intelligence to recruit & train thousands of employees. *Forbes*. https://www.forbes.com/sites/bernardmarr/2018/12/14/the-amazing-ways-how-unilever-uses-artificial-intelligence-to-recruit-train-thousands-of-employees/

McFadyen, J., Habicht, J., Dina, L.-M., Harper, R., Hauser, T. U., & Rollwage, M. (2024). AI-enabled conversational agent increases engagement with cognitive-behavioral therapy: A randomized controlled trial. MEDRXIV. https://doi.org/10.1101/2024.11.01.24316565

Mehrabi, N., Morstatter, F., Saxena, N., Lerman, K., & Galstyan, A. (2021). A survey on bias and fairness in machine learning. ACM Computing Surveys, 54(6), 1-35. https://doi.org/10.1145/3457607

Mitchell, S., Potash, E., Barocas, S., D'Amour, A., & Lum, K. (2021). Algorithmic fairness: Choices, assumptions, and definitions. *Annual Review of Statistics and Its Application*, 8, 141–163. https://doi.org/10.1146/annurev-statistics-042720-125902

Mittelstadt, B. D., Allo, P., Taddeo, M., Wachter, S., & Floridi, L. (2016). The Ethics of Algorithms: Mapping the Debate. *Big Data & Society,* 3(2), 2053951716679679. https://doi.org/10.1177/2053951716679679

Mittelstadt, B. D. (2019). Principles alone cannot guarantee ethical AI. *Nature Machine Intelligence*, 1(11), 501–507. https://doi.org/10.1038/s42256-019-0114-4

Mohammad Amini, M., Jesus, M., Fanaei Sheikholeslami, D., Alves, P., Hassanzadeh Benam, A., & Hariri, F. (2023). Artificial Intelligence Ethics and Challenges in Healthcare Applications: A Comprehensive Review in the Context of the European GDPR Mandate. *Machine Learning and Knowledge Extraction*, 5(3), 1023-1035. https://doi.org/10.3390/make5030053

Mount Sinai Health System. (2024, September 12). *Mount Sinai Health System and IBM Research launch effort that leverages artificial intelligence and behavioral data to improve mental health care for young people.* https://www.mountsinai.org/about/newsroom/2024/mount-sinai-health-system-and-ibm-research-launch-effort-that-leverages-artificial-intelligence-and-behavioral-data-to-improve-mental-health-care-for-young-people

Mukherjee, A., & Chang, H. H. (2024). *AI knowledge and reasoning: Emulating expert creativity in scientific research* (arXiv Preprint No. 2404.04436). *arXiv.* https://arxiv.org/abs/2404.04436

National Institutes of Health. (2023). Analysis of social media language using AI models predicts depression severity for White Americans but not Black Americans. NIH News Releases. https://www.nih.gov/news-events/news-releases/

Nepal, S., Pillai, A., Wang, W., Griffin, T., Collins, A. C., Heinz, M., Lekkas, D., Mirjafari, S., Nemesure, M., Price, G., Jacobson, N. C., & Campbell, A. T. (2024). MoodCapture: Depression detection using in-the-wild smartphone images. *Proceedings of the 2024 CHI Conference on Human Factors in Computing Systems, 2024*, 996. https://doi.org/10.1145/3613904.3642680

Nikbin, S., & Qu, Y. (2024). A study on the accuracy of micro expression based deception detection with hybrid deep neural network models. *European Journal of Electrical Engineering and Computer Science, 8*(3), 14–20. https://doi.org/10.24018/ejece.2024.8.3.610

Nilsen, Per & Thor, Johan & Bender, Miriam & Leeman, Jennifer & Andersson Gäre, Boel & Sevdalis, Nick. (2022). Bridging the Silos: A Comparative Analysis of Implementation Science and Improvement Science. *Frontiers in Health Services*. 1. 10.3389/frhs.2021.817750.

Obermeyer, Z., Powers, B., Vogeli, C., & Mullainathan, S. (2019). Dissecting racial bias in an algorithm used to manage the health of populations. *Science, 366*(6464), 447-453. https://doi.org/10.1126/science.aax2342

Ogunwale, A., Smith, A., Fakorede, O., & Ogunlesi, A. O. (2024). Artificial intelligence and forensic mental health in Africa: A narrative review. *International Review of Psychiatry.* https://doi.org/10.1080/09540261.2024.2405174

Perkins, M., Furze, L., Roe, J., & MacVaugh, J. (2024*). The Artificial Intelligence Assessment Scale (AIAS): A Framework for Ethical Integration of Generative AI in Educational Assessment.* Journal of University Teaching and Learning Practice, 21(6). https://doi.org/10.53761/q3azde36

Perochon, S., Di Martino, J. M., Carpenter, K. L. H., Compton, S., Davis, N., Eichner, B., Espinosa, S., Franz, L., Krishnappa Babu, P. R., Sapiro, G., & Dawson, G. (2023). Early detection of autism using digital behavioral phenotyping. *Nature Medicine*, 29(10), 2489–2497. https://doi.org/10.1038/s41591-023-02574-3

Peterson Health Technology Institute. (2025). *Adoption of artificial intelligence in healthcare delivery systems: Early applications and impacts.* https://phti.org/resources

Pozzi, G., De Proost, M. Keeping an AI on the mental health of vulnerable populations: reflections on the potential for participatory injustice. *AI Ethics* 5, 2281–2291 (2025). https://doi.org/10.1007/s43681-024-00523-5

Pressler, S. J., Jung, M., Gradus-Pizlo, I., Titler, M. G., Smith, D. G., Gao, S., Lake, K. R., Burney, H., Clark, D. G., Wierenga, K. L., Dorsey, S. G., & Giordani, B. (2022). Randomized Controlled Trial of a Cognitive Intervention to Improve Memory in Heart Failure. *Journal of cardiac failure*, 28(4), 519–530. https://doi.org/10.1016/j.cardfail.2021.10.008

Price, W. N., II, & Cohen, I. G. (2019). Privacy in the age of medical big data. *Nature Medicine*, 25(1), 37–43. https://doi.org/10.1038/s41591-018-0272-7

Psychology Today. (2024, February). 3 ways we're already using AI in mental health care. https://www.psychologytoday.com/us/blog/a-different-kind-of-therapy/202402/3-ways-were-already-using-ai-in-mental-health-care

Quijano-Sánchez, L., Liberatore, F., Camacho-Collados, J., & Camacho-Collados, M. (2018).

Applying automatic text-based detection of deceptive language to police reports: Extracting behavioral patterns from a multi-step classification model to understand how we lie to the police. *Knowledge-Based Systems*, 149, 155–168. https://doi.org/10.1016/j.knosys.2018.03.010

Rajkomar, A., Lin, A. L., & Sundt, E. M. (2024). Generative artificial intelligence in behavioral health. *NEJM AI, 1*(3). https://doi.org/10.1056/AIoa2400802

Riordan, A., Echeverria, V., Jin, Y., Yan, L., Swiecki, Z., Gašević, D., & Martinez-Maldonado, R. (2024). Human-centred learning analytics and AI in education: A systematic literature review. *Computers and Education: Artificial Intelligence*, 6, 100215. https://doi.org/10.1016/j.caeai.2024.100215

Richardson, R., Schultz, J. M., & Crawford, K. (2019). *Dirty data, bad predictions: How civil rights violations impact police data, predictive policing systems, and justice.* New York University Law Review, 94(2), 192–233. https://www.nyulawreview.org/wp-content/uploads/2019/04/NYULawReview-94-Richardson-Schultz-Crawford.pdf

Sadeh-Sharvit, S., Camp, T. D., Horton, S. E., Hefner, J. D., Berry, J. M., Grossman, E., & Hollon, S. D. (2023). Effects of an artificial intelligence platform for behavioral interventions on depression and anxiety symptoms: Randomized clinical trial. *Journal of Medical Internet Research*, 25, e46781. https://doi.org/10.2196/46781

Saha, B., Tahora, S., Barek, A., & Shahriar, H. (2023). *HIPAAChecker: The comprehensive solution for HIPAA compliance in Android mHealth apps. arXiv.* https://arxiv.org/abs/2306.06448

Sauerbrei, A., Kerasidou, A., Lucivero, F., & Hallowell, N. (2023). The impact of artificial intelligence on the person-centred, doctor–patient relationship: Some problems and solutions. *BMC Medical Informatics and Decision Making, 23*(1), 73. https://doi.org/10.1186/s12911-023-02162-y

Savana. (2024). *NLP-as-a-Service: Turning clinical notes into real-world evidence.* Retrieved from https://savanamed.com/nlp-as-a-service/

Shirley, H. M., & Nair, B. M. (2023). The efficacy of artificial intelligence-driven Immersive Reader for dyslexic students in special schools: A case study. *Journal of English Language Teaching, 65*(5), 3–8.

Siddals, S., Torous, J. & Coxon, A. *"It happened to be the perfect thing"*: experiences of generative AI chatbots for mental health. *npj Mental Health Res* **3**, 48 (2024). https://doi.org/10.1038/s44184-024-00097-4

Squassina, A., Paribello, P., Pinna, M., Contu, M., Pisanu, C., Congiu, D., Severino, G., Meloni, A., Carta, A., Conversano, C., Mola, F., Del Zompo, M., Bernoni d'Aversa, F., Minelli, A., Gennarelli, M., Pinna, F., Carpiniello, B., & Manchia, M. (2025). A naturalistic retrospective evaluation of the utility of pharmacogenetic testing based on CYP2D6 and CYP2C19 profiling in antidepressants treatment in a cohort of patients with major depressive disorder. *Progress in Neuropsychopharmacology & Biological Psychiatry, 137,* 111292. https://doi.org/10.1016/j.pnpbp.2025.111292

Sreedharan, S. (2023). Human-aware AI: A foundational framework for human–AI interaction. *AI Magazine, 44*(4), 460–466. https://doi.org/10.1002/aaai.12142

Stackpole, T. (2021, June 4). Inside IKEA's digital transformation. *Harvard Business Review.* https://hbr.org/2021/06/inside-ikeas-digital-transformation

Stade, E. C., Eichstaedt, J. C., Kim, J. P., & Wiltsey Stirman, S. (2025). Readiness evaluation for artificial intelligence–mental health deployment and implementation (READI): A review and proposed framework. *Technology, Mind, and Behavior. Advance online publication.* https://doi.org/10.1037/tmb0000163

Starke, G., D'Imperio, A., & Ienca, M. (2023). Out of their minds? Externalist challenges for using AI in forensic psychiatry. *Frontiers in Psychiatry, 14,* 1209862. https://doi.org/10.3389/fpsyt.2023.1209862

Staton, B. (2024, November 7). Employers look to AI tools to plug skills gap and retain staff. *Financial Times*. https://www.ft.com/content/9cf58a76-5245-4cdf-9449-239e90077eb5

Stein, N., & Brooks, K. (2017). A fully automated conversational artificial intelligence for weight loss: Longitudinal observational study among overweight and obese adults. *JMIR Diabetes*, 2(2), e28. https://doi.org/10.2196/diabetes.8590

Sterz, S., Baum, K., Biewer, S., Hermanns, H., Lauber-Rönsberg, A., Meinel, P., & Langer, M. (2024). *On the quest for effectiveness in human oversight: Interdisciplinary perspectives*. ACM Conference on Fairness, Accountability, and Transparency (FAccT). https://doi.org/10.1145/3630106.3659051

Straw, I., & Callison-Burch, C. (2020). Artificial intelligence in mental health and the biases of language-based models. *PLOS ONE*, 15(12), e0240376. https://doi.org/10.1371/journal.pone.0240376

Su, C., Xu, Z., Pathak, J., & Wang, F. (2020). Deep learning in mental health outcome research: A scoping review. *Translational Psychiatry*, 10(1), 116. https://doi.org/10.1038/s41398-020-0780-3

Tang, H., Miri Rekavandi, A., Rooprai, D., & others. (2024). Analysis and evaluation of explainable artificial intelligence on suicide risk assessment. *Scientific Reports*, *14,* 6163. https://doi.org/10.1038/s41598-024-53426-0

Tavory, T. (2024). Regulating AI in mental health: Ethics of care perspective. *JMIR Mental Health*, 11, e58493. https://doi.org/10.2196/58493

Thakkar, A., Gupta, A., & De Sousa, A. (2024). Artificial intelligence in positive mental health: A narrative review. FRONTIERS IN DIGITAL HEALTH, 6, 1280235. https://doi.org/10.3389/fdgth.2024.1280235

Timmons, A. C., Duong, J. B., Simo Fiallo, N., Lee, T., Vo, H. P. Q., Ahle, M. W., Comer, J. S., Brewer, L. C., Frazier, S. L., & Chaspari, T. (2023). A call to action on assessing and mitigating bias in artificial intelligence applications for mental health. *Perspectives on Psychological Science*, 18(5), 1062–1096. https://doi.org/10.1177/17456916221134490

Tong, Y., Patel, M., Kording, K., & Lindquist, K. A. (2024). *A large-scale analysis of causal knowledge in psychology using GPT-4*. arXiv preprint arXiv:2402.14424. https://doi.org/10.48550/arXiv.2402.14424

Topol, E. (2019). *Deep medicine: How artificial intelligence can make healthcare human again*. Basic Books.

Torous, J., Bucci, S., Bell, I. H., Kessing, L. V., Faurholt-Jepsen, M., Whelan, P., ... & Firth, J. (2021). The growing field of digital psychiatry: Current evidence and the future of apps, social media, chatbots, and virtual reality. WORLD PSYCHIATRY, 20(3), 318–335. https://doi.org/10.1002/wps.20883

Tortora, L. (2024). Beyond discrimination: Generative AI applications and ethical challenges in forensic psychiatry. *Frontiers in Psychiatry, 15,* 1346059. https://doi.org/10.3389/fpsyt.2024.1346059

Tortora, L., Meynen, G., Bijlsma, J., Tronci, E., & Ferracuti, S. (2020). Neuroprediction and A.I. in forensic psychiatry and criminal justice: A neurolaw perspective. *Frontiers in Psychology*, 11, 220. https://doi.org/10.3389/fpsyg.2020.00220

Tsirmpas, C., Nikolakopoulou, M., Kaplow, S., Andrikopoulos, D., Fatouros, P., Kontoangelos, K., & Papageorgiou, C. (2023). A digital mental health support program for depression and anxiety in populations with attention-deficit/hyperactivity disorder: Feasibility and usability study. *JMIR Formative Research*, 7, e48362. https://doi.org/10.2196/48362

Ulfert, A.-S., Le Blanc, P., González-Romá, V., Grote, G., & Langer, M. (2024). Are we ahead of the trend or just following? The role of work and organizational psychology in shaping emerging technologies at work. *European Journal of Work and Organizational Psychology, 33*(2), 120–129. https://doi.org/10.1080/1359432X.2024.2324934

Umashankar, N., & Geethanjali, K. (2024). *From stress to support: An AI-powered chatbot for student mental health care*. engrXiv. https://engrxiv.org/preprint/view/4156/version/5681

Um, J., Park, J., Lee, D. E., Ahn, J. E., & Baek, J. H. (2025). Machine learning models to identify individuals with imminent suicide risk using a wearable device: A pilot study. Psychiatry Investigation, 22(2), 156–166. https://doi.org/10.30773/pi.2024.0257

U.S. Department of Health & Human Services. (n.d.). *Summary of the HIPAA Privacy Rule*. HHS.gov. https://www.hhs.gov/hipaa/for-professionals/privacy/laws-regulations/index.html

Wall, D. P., Dally, R., Luyster, R., Jung, J.-Y., & DeLuca, T. F. (2012). Use of artificial intelligence to shorten the behavioral diagnosis of autism. *PLOS ONE*, 7(8), e43855. https://doi.org/10.1371/journal.pone.0043855

Wang, H., & Avillach, P. (2021). Diagnostic classification and prognostic prediction using common genetic variants in autism spectrum disorder: Genotype-based deep learning. *JMIR Medical* Informatics, 9(4), e24754. https://doi.org/10.2196/24754

White House Office of Science and Technology Policy. (2022*). Blueprint for an AI Bill of Rights: Making automated systems work for the American people.* Retrieved from https://bidenwhitehouse.archives.gov/ostp/ai-bill-of-rights/

World Health Organization. (2021). *Mental health atlas 2020*. https://www.who.int/publications/i/item/9789240036703 World Health Organization. (2021).

Zhang, W., Liu, X., Wang, L., & Huang, Y. (2025). Scaling up personalized digital mental health care with AI: Opportunities and challenges. BMC PSYCHIATRY, 25, 64. https://doi.org/10.1186/s12888-025-06483-2

Zhang, Y., Folarin, A. A., Dineley, J., Conde, P., de Angel, V., Sun, S., Ranjan, Y., Rashid, Z., Stewart, C., Laiou, P., Sankesara, H., Qian, L., Matcham, F., White, K., Oetzmann, C., Lamers, F., Siddi, S., Simblett, S., ... Cummins, N. (2023). *Identifying depression-related topics in smartphone-collected free-response speech recordings using an automatic speech recognition system and a deep learning topic model.* Preprint at arXiv:2308.11773. https://doi.org/10.48550/arXiv.2308.11773

Zhang, Z., & Wang, J. (2024). Can AI replace psychotherapists? Exploring the future of mental health care. Frontiers in Psychiatry, 15, 1444382. https://doi.org/10.3389/fpsyt.2024.1444382

Zheng, L., Wang, O., Hao, S., Ye, C., Liu, M., Xia, M., et al. (2020). Development of an early-warning system for high-risk patients for suicide attempt using deep learning and electronic health records. *Translational Psychiatry*, 10, Article 72. https://doi.org/10.1038/s41398-020-0684-2

Zhou, S., Zhao, J., & Zhang, L. (2022). Application of artificial intelligence on psychological interventions and diagnosis: An overview. *Frontiers in Psychiatry, 13*, 811665. https://doi.org/10.3389/fpsyt.2022.811665

Zilcha-Mano, S., Zhu, X., Suarez-Jimenez, B., Pickover, A., Tal, S., Such, S., Marohasy, C., Chrisanthopoulos, M., Salzman, C., Lazarov, A., Neria, Y., & Rutherford, B. R. (2020). Diagnostic and predictive neuroimaging biomarkers for posttraumatic stress disorder. *Biological Psychiatry: Cognitive Neuroscience and Neuroimaging, 5(7)*, 688–696. https://doi.org/10.1016/j.bpsc.2020.03.010

Zilka, T., Harari, Y., & Barash, E. (2023). *The progression of disparities within the criminal justice system: Differential enforcement and risk assessment instruments. arXiv.* https://arxiv.org/abs/2305.07575

Zolnoori, M., Vergez, S., Xu, Z., Esmaeili, E., Zolnour, A., Briggs, K. A., Scroggins, J. K., Hosseini Ebrahimabad, S. F., Noble, J. M., Topaz, M., Bakken, S., Bowles, K. H., Spens, I., Onorato, N., Sridharan, S., & McDonald, M. V. (2024). Decoding disparities: Evaluating automatic speech recognition system performance in transcribing Black and White patient verbal communication with nurses in home healthcare. *JAMIA Open*, 7(4), ooae130. https://doi.org/10.1093/jamiaopen/ooae130

Glossary of Terms

A

- **Artificial Intelligence (AI)** – A branch of computer science focused on creating systems capable of performing tasks that typically require human intelligence, such as learning, reasoning, problem-solving, and language comprehension.
- **Algorithm** – A step-by-step set of rules or calculations followed by a computer to perform a specific task.
- **Artificial Neural Network (ANN)** – A type of AI model inspired by the structure and functioning of the human brain, used for pattern recognition, decision-making, and data analysis.
- **Automation Bias** – The human tendency to overly trust AI-generated output—even when it may be inaccurate or inappropriate, especially if it appears confident or professional.

B

- **Bias in AI** – Systematic errors in AI models caused by imbalanced or unrepresentative training data, leading to unfair outcomes in psychological assessments or decision-making.
- **Bias Audit** – An evaluation process used to test AI models for performance disparities across demographic groups, identifying potential inequities in predictions or outcomes.
- **Big Data** – Large and complex datasets that AI systems analyze to identify patterns, trends, and insights in psychological research and practice.
- **Behavioral Data** – Information collected about an individual's actions, such as speech patterns, facial expressions, or online interactions, used in AI-driven psychological assessments.

C

- **Chatbot** – An AI-driven conversational agent that interacts with users through text or speech, often used in mental health support and therapy applications.
- **Cognitive Behavioral Therapy (CBT)** – A structured, evidence-based psychological treatment that helps individuals identify and change negative thought patterns and behaviors.
- **Computer Vision** – A subset of AI that enables machines to analyze and interpret visual information, such as facial expressions and gestures, for psychological assessments.

- **Conversational Agent** – A type of chatbot or virtual assistant that engages in dialogue using natural language processing, often used in mental health apps or digital therapy tools.
- **Clinical Decision Support System (CDSS)** – AI-powered tools that assist psychologists and mental health professionals in making evidence-based clinical decisions.

D

- **Deep Learning** – A subset of machine learning that uses multi-layered neural networks to analyze complex data and improve AI decision-making.
- **Data Privacy** – Regulations and practices that ensure personal and psychological data are securely stored, processed, and protected from unauthorized access.
- **Decision Support Systems (DSS)** – AI-based tools designed to assist mental health professionals in analyzing patient data and making informed treatment decisions.
- **Digital Twin (Psychological Context)** – A simulated model designed to mirror the behavior, personality traits, or psychological profile of a specific individual, often used in AI-driven personalization or training.

E

- **Ethical AI** – The development and use of AI technologies that prioritize fairness, transparency, privacy, and human well-being in psychological practice.
- **Explainability** – The ability of an AI system to provide understandable and transparent reasoning behind its decisions and predictions.
- **Explainable AI (XAI)** – AI systems designed to provide understandable reasons for their outputs—helping users see what factors influenced a prediction or decision.

F

- **Forensic Psychology** – The application of psychological principles in legal and criminal justice settings, including AI-assisted risk assessments and deception detection.
- **Fairness in AI** – The principle of ensuring that AI models do not reinforce discrimination or bias in psychological assessments and treatment recommendations.

G

- **General AI** – A hypothetical form of AI that would have human-like cognitive abilities, capable of performing any intellectual task that a human can do.
- **Generative AI** – AI models that create new content, such as text, images, and speech, used in therapy simulations and psychological research.

- **Ground-Truth Labels** – The correct, verified outcomes or classifications used to train and test machine learning models. In psychological applications, this might refer to clinician-confirmed diagnoses or validated assessment scores.

H

- **Hallucination (in AI)** – The tendency of generative models to produce information that sounds plausible but is false, fabricated, or not grounded in the input data.
- **Human-in-the-Loop (HITL)** – A model of AI deployment where humans are actively involved in reviewing, guiding, or overriding AI outputs to ensure safety, accuracy, and ethical use.

I

- **Industrial-Organizational Psychology** – The study of human behavior in the workplace, where AI is used for hiring assessments, leadership development, and employee well-being monitoring.
- **Intelligent Virtual Assistant (IVA)** – AI-powered digital assistants that provide mental health support and therapy recommendations.

L

- **Language Model** – A type of AI designed to understand and generate human language, used in psychological chatbots and sentiment analysis.
- **Large Language Model (LLM)** – A powerful AI system trained on extensive text datasets to process and generate human-like responses in psychological applications.
- **Learning Algorithm** – A mathematical model that allows AI to improve its performance over time by analyzing new data.

M

- **Machine Learning (ML)** – A subset of AI that enables systems to learn from data without being explicitly programmed, widely used in psychological assessments and mental health analytics.
- **Mental Health Informatics** – The application of AI and data science in mental health care, including predictive modeling and digital therapy tools.
- **Model Bias** – A systematic error in predictions resulting from training data that underrepresents or misrepresents certain populations or contexts.
- **Model Drift** – A phenomenon where an AI model's performance declines over time due to changes in data patterns, environments, or user behavior—especially relevant in dynamic clinical or educational settings.
- **Multimodal AI** – AI systems that process and integrate information from multiple data types—such as text, voice, facial expressions, and biometric

data—to generate a more comprehensive understanding of behavior or emotion.

N

- **Natural Language Processing (NLP)** – A field of AI that focuses on enabling computers to understand, interpret, and generate human language, used in therapy chatbots and sentiment analysis.

- **Neural Network** – A computational model inspired by the human brain, used in deep learning to recognize patterns in psychological and behavioral data.

O

- **Overfitting** – When an AI model performs well on its training data but poorly on new or diverse data because it has memorized rather than generalized patterns.

P

- **Predictive Analytics** – AI techniques used to forecast psychological outcomes, such as the likelihood of depression relapse or treatment success.
- **Personalized AI Therapy** – AI-driven mental health interventions tailored to individual patient needs based on behavioral data and previous therapy sessions.

R

- **Reinforcement Learning (RL)** – A machine learning approach in which AI models learn by receiving rewards or penalties based on their actions, used in adaptive therapy tools.
- **Retrieval-Augmented Generation (RAG)** – A method that enhances language model responses by retrieving relevant external information (e.g., clinical notes, client history) during generation, making outputs more personalized and context-aware.

S

- **Sentiment Analysis** – An NLP technique that evaluates emotional tone in speech or text, used in AI-powered mental health applications.
- **Supervised Learning** – A machine learning method in which AI is trained on labeled data to recognize patterns and make predictions in psychological research.
- **Speech Emotion Recognition (SER)** – AI-powered analysis of vocal tone, pitch, and rhythm to detect emotional states in mental health assessments.

T

- **Transparency in AI** – The principle that AI systems should provide clear, interpretable explanations for their decisions in psychological practice..

U

- **Unsupervised Learning** – A machine learning technique in which AI identifies hidden patterns in data without predefined labels, useful in personality trait analysis and forensic profiling.

W

- **Wearable AI** – Smart devices that use AI to track physiological and psychological indicators, such as heart rate variability and stress levels, for mental health monitoring.

www.ingramcontent.com/pod-product-compliance
Lightning Source LLC
Chambersburg PA
CBHW041609260326
41914CB00012B/1433